While there are many excellent works on climate diplomacy that also look into the growing role of the emerging economies of India, China, Brazil and South Africa, this book adopts a particularly useful and novel perspective to analyse, collectively and individually, their climate diplomacy positions. By focusing on the key role played by 'ideas' in influencing the climate debates rather than simply taking the 'international climate order' as a given, it provides a uniquely multidimensional exposition of global climate politics from these countries' perspective through 'formula narratives'. The case study of India, in particular, provides a deep understanding of the interaction between various drivers, including ones arising from domestic and international imperatives.

— Cleo Paskal, associate fellow in the Energy, Environment and Resources programme and the Asia-Pacific programme, Chatham House.

Climate Diplomacy and Emerging Economies

This book analyses the role of the BASIC countries – Brazil, South Africa, India, and China – in the international climate order. *Climate Diplomacy and Emerging Economies* explores the collective and individual positions of these countries towards climate diplomacy, focusing in particular on the time period between the 2009 and 2019 climate summits in Copenhagen and Madrid. Dhanasree Jayaram examines the key drivers behind their climate-related policies (both domestic and international) and explores the contributory role of ideational and material factors (and the interaction between them) in shaping the climate diplomacy agenda at multilateral, bilateral, and other levels. Digging deeper into the case study of India, Jayaram studies the shifts in its climate diplomacy by looking into the ways in which climate change is framed and analyses the variations in perceptions of the causes of climate change, the solutions to it, the motivations for setting climate action goals, and the methods to achieve the goals.

This book will be of great interest to students and scholars of climate change, environmental policy, and politics and IR more broadly.

Dhanasree Jayaram is currently Assistant Professor in the Department of Geopolitics and International Relations and Co-coordinator at the Centre for Climate Studies, Manipal Academy of Higher Education (MAHE), Karnataka, India. She is also Research Fellow, Earth System Governance (ESG). Dr Jayaram is a member of the Climate Security Expert Network (CSEN), supported by a grant from the German Federal Foreign Office and whose Secretariat is run by the Berlin-based think tank adelphi. She is the author of *Breaking out of the Green House: India's Leadership in Times of Environmental Change* (2012).

Routledge Focus on Environment and Sustainability

Traditional Ecological Knowledge in Georgia
A Short History of the Caucasus
Zaal Kikvidze

Traditional Ecological Knowledge and Global Pandemics
Biodiversity and Planetary Health Beyond Covid-19
Ngozi Finette Unuigbe

Climate Diplomacy and Emerging Economies
India as a Case Study
Dhanasree Jayaram

Linking the European Union Emissions Trading System
Political Drivers and Barriers
Charlotte Unger

Post-Pandemic Sustainable Tourism Management
The New Reality of Managing Ethical and Responsible Tourism
Tony O'Rourke and Marko Koščak

Consumption Corridors
Living a Good Life within Sustainable Limits
Doris Fuchs, Marlyne Sahakian, Tobias Gumbert, Antonietta Di Giulio, Michael Maniates, Sylvia Lorek and Antonia Graf

For more information about this series, please visit: www.routledge.com/
Routledge-Focus-on-Environment-and-Sustainability/book-series/RFES

Climate Diplomacy and Emerging Economies
India as a Case Study

Dhanasree Jayaram

First published 2021
by Routledge
2 Park Square, Milton Park, Abingdon, Oxon OX14 4RN

and by Routledge
52 Vanderbilt Avenue, New York, NY 10017

Routledge is an imprint of the Taylor & Francis Group, an informa business

© 2021 Dhanasree Jayaram

The right of **Dhanasree Jayaram** to be identified as author of this
work has been asserted by her in accordance with sections 77 and 78
of the Copyright, Designs and Patents Act 1988.

British Library Cataloguing-in-Publication Data
A catalogue record for this book is available from the British Library

Library of Congress Cataloging-in-Publication Data
A catalog record has been requested for this book

ISBN: 978-0-367-63402-5 (hbk)
ISBN: 978-0-367-63404-9 (pbk)
ISBN: 978-1-003-11907-4 (ebk)

Typeset in Times New Roman
by codeMantra

Contents

Figures

Tables

Acknowledgements

This project would not have materialised without the support of many individuals, institutions, and networks. I would like to first thank Annabelle Harris, Editor – Environment and Sustainability (Routledge) and Oindrila Bose, Editorial Assistant, for being so patient with me throughout the publication process. I also would like to thank the blind peer-reviewers who went through my book proposal and gave me constructive suggestions to improve the manuscript's quality.

I would like to extend my gratitude to the Federal Commission for Scholarships for Foreign Students (FCS), Government of Switzerland, for offering me the generous Swiss Government Excellence Scholarship in 2018 to work on this project as a part of my postdoctoral research. In this respect, I would like to thank my host, the University of Lausanne, and my supervisor, Dr Yohan Ariffin, Associate Professor, Faculty of Social and Political Sciences, for their immense contribution to this project. A large part of the study's conceptualisation, methodology, and analysis was facilitated through fruitful conversations with Prof. Ariffin. The other faculty, students, and staff at the university were equally helpful in helping me settle into the new place and work on this project successfully. Here, I would like to specially mention Dr Lucile Maertens and Dr Chanatporn Limprapoowiwattana for being the best support system that I could have imagined.

I would fail at my duty if I did not thank Manipal Academy of Higher Education (MAHE) and its various departments and centres that I am a part of. This institution has given me all the space and liberty to pursue my goals in academia right from the beginning. I would like to thank my superiors and colleagues at the Department of Geopolitics & International Relations, Centre for Climate Studies and Manipal Advanced Research Group for being highly supportive throughout my postdoctoral studies. Besides the leadership of the institution, I would like to specifically mention one individual's name – Prof. M. D. Nalapat – who has always remained the beacon of hope when the going gets tough for me, and who also kindly agreed to write the foreword for this book.

I would like to thank all the interviewees who agreed to interact with me through various means, especially in COVID-19 times. The number of interactions I have had with experts and scholars on this theme is endless in the past two years. While I would have loved to name all of them, I am more fearful of missing a few names, which would be totally unfair. However, here too, I need to mention adelphi (with whom I have been working on the project, "Climate Diplomacy") and the Earth System Governance for influencing this study, both directly and indirectly. My friends, peers, and mentors in these networks and organisations have been instrumental in many respects – from connecting me to experts to providing research inputs for the study.

Last but not the least, my family has always been my backbone. This book would not have been completed but for their constant motivation. My friends – Ramu, Nadeem, and Fatima – helped me a lot with finalising several parts of the manuscript. I cannot thank them enough for their unstinting assistance in this regard.

I would also like to acknowledge that a few parts of this manuscript are based on some of my previously published articles and one forthcoming article:

- Jayaram, Dhanasree. Forthcoming. "India's Climate Diplomacy towards the EU: From Copenhagen to Paris and Beyond." In *EU-India Relations: The Strategic Partnership in the Light of the European Union Global Strategy*, edited by Philipp Gieg, Timo Lowinger, Manuel Pietzko, Anja Zürn, Ummu Salma Bava and Gisela Müller-Brandeck-Bocquet. Cham: Springer.
- Belis, David, Schunz, Simon, Wang, Tao and Jayaram, Dhanasree. 2018. "Climate Diplomacy and the Rise of 'Multiple Bilateralism' between China, India and the EU." *Carbon & Climate Law Review* 12 (2): 85–97.
- Jayaram, Dhanasree. 2018. "From 'Spoiler' to 'Bridging Nation': The Reshaping of India's Climate Diplomacy." *La Revue internationale et stratégique* 109: 181–190.
- Jayaram, Dhanasree. 2015. "A Shift in the Agenda for China and India: Geopolitical Implications for Future Climate Governance." *Carbon & Climate Law Review* 9 (3): 219–230.
- Jayaram, Dhanasree. 2014. "Sino-Indian Cooperation at the Climate Change Negotiations: The Past, the Present and the Future." *Chinese Journal of Urban and Environmental Studies* 2 (1): 1450010.

Foreword

Emerging economies have played a crucial role in the international climate order since the beginning of the climate change negotiations. However, their role became more prominent since the 2009 Copenhagen Summit, when India, China, Brazil, and South Africa resisted pressure exerted by historically polluting countries to take on one-sided emission reduction commitments. The positions taken by them have varied over a period of time, up until the Paris Agreement was signed and even beyond.

In this context, Dr Dhanasree Jayaram's study – "Climate Diplomacy and Emerging Economies: India as a Case Study" – analyses the role of the emerging economies in the international climate order. Under this, the BASIC countries and the contributory role of ideational and material factors (and the interaction between them) in shaping their climate diplomacy during 2009–2019 are studied in depth. The book uses the case study of India and analyses the shifts in its climate diplomacy during 2009–2019 by looking into its framing of climate change and the use of certain formula narratives based on ideas of equity, climate justice, sovereignty, leadership, and so on.

Dr Jayaram's narratives identify and analyse the variations in perceptions of the causes of climate change, the solutions to it, the motivations for setting climate action goals, and the methods to achieve the goals. The study goes deeper into the drivers of the emergence of the BASIC countries at the international level (such as the 2007–2008 financial crisis and aspirations of global climate leadership). However, it does not stop at identifying the "solidarity" among the BASIC countries. It goes beyond to identify the divisions among them owing to various reasons, including the understanding reached between China and the US in 2014 and South Africa's identification with its regional identity (more than with emerging economies elsewhere).

The book analyses the climate diplomacy positions of China, Brazil, and South Africa by looking into both the domestic dynamics and their interlinkages with the international positioning. From elevating the issue of climate change to the level of State Council in 2007 to becoming the global leader in clean/green technology (mainly renewable energy), China's journey in the climate change negotiations has been shaped by its desire to be seen in the league of great powers as well as its domestic compulsion of tackling pollution. Lula's Brazil, even while struggling to manage the deteriorating deforestation rates, evolved as a potential climate leader with its investments in green fuels (ethanol) domestically and measures to ensure equitable distribution of the carbon space at the international level. South Africa, a coal-dependent economy, has combined its notions of equity with the need to usher in reforms through measures such as carbon tax.

These countries and India have been demanding financial and technological assistance from the rich world owing to the latter's historical responsibility towards causing climate change in the first place. However, over time such demands have given way to a more pragmatic and result-oriented approach, such as building national capacities through bilateral partnerships.

The study highlights the case study of India as representative of emerging economies – showing the relevance of the above-mentioned ideas and their interrelationships with material interests. Not only is a historical background to India's positions at the negotiations given, an attempt is also made to analyse the evolution of its bilateral climate diplomacy initiatives with the other countries/parties, including the EU, China, and the US. The arguments elucidate the nuanced shifts made by India in the realm of climate diplomacy due to various factors such as advancing foreign policy interests, enhancing its own clean energy base and promoting energy security through these efforts, mitigating the increasing climate vulnerabilities, and so on. Although India continues to uphold the traditionally held positions on equity and climate justice, it has since Narendra Modi took charge as Prime Minister become more flexible on these fronts. This has caused it to be seen as a part of the solution, despite having one of the lowest per capita emissions. The launch of the International Solar Alliance, which is the brainchild of Prime Minister Modi, reflects the change in mood.

The framing of climate change as a vulnerability/crisis/challenge, an opportunity (in terms of economic and social co-benefits domestically and that of leadership internationally), a development issue (issues related to poverty, trade-offs between environment and development, and so on), a geopolitical construct (power hierarchies, coalitions, and alignments), and a cultural norm (with human interest, accommodating civilizational and cultural values) is dealt with in the context of analysing climate diplomacy as a part of India's larger foreign policy agenda. Given the essentiality of such diplomacy in the future, this is a work that would be of immense value in an understanding of the evolution of climate diplomacy and its effects on policy.

— M. D. Nalapat

UNESCO Peace Chair and Vice-Chair, Manipal Advanced Research Group (MARG)

Professor and Director, Department of Geopolitics & International Relations

Manipal Academy of Higher Education

Editorial Director, The Sunday Guardian and ITV network (India)

1 Introduction

Climate diplomacy is conceptualised and practised through multiple approaches. In International Relations, it is often conflated and interlinked with ideas of cooperation and multilateralism. When seen through the prism of foreign policy, climate diplomacy becomes integral to the process of identifying entry points for climate cooperation on various issues such as energy and land use among others. The interconnectedness between climate change and foreign policy can be expressed through climate diplomacy (adelphi, n.d.):

> Climate change is one of the greatest challenges of the 21st century – the repercussions for our foreign policy agenda are substantial. A stronger role for foreign policy in international climate policy has been called for – namely through climate diplomacy.

Climate diplomacy can be considered a reflection of the persisting geopolitical scenario or it may also be seen as an instrument of geopolitics – a practice that influences the latter. Not only does climate diplomacy provide a framework for analysing geopolitics (such as North–South relations), but it is also an advocacy tool that is used by nation states and other actors (non-state and sub-state) to promote dialogue, stakeholder engagement, and so on. A good example of interplay between climate diplomacy and geopolitics is the way in which the emerging economies – namely, Brazil, South Africa, India, and China – asserted themselves both collectively and individually at the 2009 Copenhagen Summit to resist pressure from the industrialised countries to dilute differentiation between the developed and developing countries. It is therefore emphasised that they concretised their place in the global climate order as key agenda-setters and norm-creators.

Both ideational and material forces are known to guide climate diplomacy. While the 'ideational' is commonly related to constructivism, 'materiality' is linked to realism in International Relations. Some would argue that ideas and values matter more in the conceptualisation and practice of climate diplomacy, while others would argue that power politics and material forces are the sole drivers of a country's climate diplomacy. Not only does this vary from context to context, but also the dichotomous relationship between the two itself is questionable. In fact, constructivism as an analytical and methodological approach gives space for the material and ideational to be "complexly interwoven and interdependent" (Pettenger 2007, p. 6). Moreover, 'realist constructivism' is an approach that finds mainstream constructivism compatible with classical realism (Barkin 2010). This inherent interplay between constructivism and realism is more evident in the case of the emerging economies for whom ideas and discursive power are of immense importance in projecting their positions on climate change, in conjunction with 'power', at the national and international levels.

The behavioural patterns of nation states in the international climate order are governed, to a great extent, by attitudes towards issues concerning global climate governance, which are in turn shaped by perceptions and ideas such as climate justice, national interest, vulnerability/danger, opportunity, historical responsibility, leadership, sustainable development, and so on. Hence, the gradual shifts in climate diplomacy positions of the emerging economies are clearly a reflection of a change in their perceptions of issues surrounding these ideas, coupled with the influence of material forces that emanate from an expansion of their 'green' capabilities and power positioning in the international system among other shifts. These ideas can be broadly categorised into moral/ethical, environmental/ecological, socioeconomic and technology, cultural and political (including geopolitical). Climate diplomacy, in short, as an embodiment of power politics or a demonstration of international cooperation, is grounded on ideas that are critical for framing of climate change as well as issues that are linked to it.

The relevance and contours of climate diplomacy

While there are no strict definitions of climate diplomacy, the following definition sums up its nature and goals to some extent, as well as the actors that are typically involved in the process (Mabey, Gallagher and Born 2013, p. 14):

> Climate diplomacy is the interface between national interest debates and international cooperation. It is the process through which nation states – and increasingly non-governmental and substate actors – determine and work to deliver their international objectives.

Climate diplomacy is often seen through the prism of the international climate regime, which is essentially encapsulated in the United Nations Framework Convention on Climate Change (UNFCCC) and its various outcomes, including the Kyoto Protocol and Paris Agreement. However, in recent times, climate diplomacy has become a centrepiece of various other forums (such as G-20 and regional organisations) as well as bilateral relations between countries. These forums and exchanges may run parallel to the climate change negotiations, and they may also reinforce the international climate regime through iterative or other types of processes. For instance, they may be involved in implementing climate action at various levels of governance in the international system. They may also help develop trust between various parties on the future of climate action, and the climate order itself.

Climate diplomacy has been analysed through different analytical frameworks in the existing literature. For instance, the approach of 'multiple bilateralism' is defined as "a strategy that entails the maintenance of several – cooperative or confrontational – bilateral diplomatic relationships in parallel as a subset of a multilateral negotiation setting in order to reach policy objectives at that multilateral level" (Belis et al. 2018, p. 86). According to this approach, a country, while engaging in bilateral relations with multiple countries, takes into consideration its individual bilateral relations with each of them. These relations could be cooperative or adversarial in nature. Belis et al. (2015) attribute the proactive momentum on climate action from the Copenhagen Summit to Paris Summit to growing multiple bilateralism among players such as the United States (US), China, and the European Union (EU) to some extent. This could be in the form of the joint announcement reached by the US and China in 2014 in the run-up to the 2015 Paris Summit (explained further in the subsequent chapters) or the growing China–EU understanding on climate issues through discussions on energy or emissions trading (Jayaram *forthcoming*).[1]

Climate diplomacy also drives formation of alliances, coalitions, and groupings based on ideational and material forces, including specific vulnerabilities, capabilities, and interests. This is where the salience of the BASIC grouping – formed by the emerging economies of India, China, Brazil, and South Africa – becomes apparent. These countries have been steadfastly upholding the principles of Common but Differentiated Responsibilities and Respective Capabilities (CBDR-RC), right to develop (of the developing countries), historical responsibility (of the developed countries) etc. (Hurrel and Sengupta 2012).

Climate diplomacy, as Elliott (2013) points out, "is not divorced from the competing political interests that underpin it". These interests are conditioned/shaped by not just domestic factors but also geopolitical ones. As much as nation states defend their "historical and national positions", they may also use climate diplomacy to achieve their geopolitical interests, as is further explored in this study too. For instance, the emerging economies are known to position themselves as responsible players or global leaders in the arena of international climate governance (Karlsson et al. 2011). What needs to be noted here is the autonomous and intermediary roles played by the ideational and material forces that feed into these positions. The ideational and material factors are not customarily or paradigmatically disengaged from one another or they may not share any form of hierarchical relationship by which one set of ideas predominate the others. This is explained further in the subsequent sections of the chapter. Therefore, as perceptions surrounding climate change and solutions evolve and shape countries' climate diplomacy, it also facilitates material progress by uncovering and creating new opportunities in the economic, social, technological, financial, and strategic domains among others (Hsu et al. 2015; Renner 2015).

The interlinkages between power hierarchies within the international system and climate diplomacy cannot be ignored while dissecting the latter. From the emergence of G-2 (the US and China), signifying a certain geopolitical realignment (Bush 2011), to China and India projecting themselves as leaders of the newly emerged climate order in the aftermath of the US's withdrawal from the Paris Agreement under the Donald Trump Presidency can be regarded as a combination of ideas and material forces feeding into their climate diplomacy positions (Zhang et al. 2017). Climate diplomacy is also used as a tool for soft power projection. Nation states that possess certain capabilities can offer their resources (through grant, aid, capacity-building etc.) to smaller and resource-poor countries for strategic purposes by winning their goodwill and confidence (Karakır 2018).

Climate diplomacy is, without a second thought, influenced by domestic interests. With climate change being mainstreamed into various policy areas (such as energy, urbanisation, food, and water), the range of actors has increased to include transnational and private/civil society organisations as well. Apart from the head of the state whose special interest in climate change could prove to be consequential, other actors, including governmental agencies, ministries, nongovernmental organisations, businesses, and media, also play a role in framing climate change–related issues domestically and internationally. With the Paris Agreement opening up the multilateral process and the realm of international climate policy in general much more for the participation and contribution of sub-state and non-state actors, these actors' stakes and responsibilities have become more important to discuss. Non-state actors, for instance, are engaged not only in "activist efforts" aimed at gaining attention but also in "climate governance" in the form of "transnational networks, epistemic communities, public–private partnerships (PPPs), and multi-stakeholder partnerships" (Kuyper, Linnér and Schroe 2017). In effect, the international climate order or regime is not a monolith; it is rather an archetype of multilevel, networked governance, involving multiple actors with varied interests, ideas, and capacities (Jayaram *forthcoming*).

Relevance of the emerging economies in the international climate order

Although the emerging economies' climate diplomacy positions became the centre of attention at the Copenhagen Summit, their contribution to shaping the global debates on climate change has been highly influential since the beginning of talks on tackling climate change, even before the UNFCCC was established in 1992. Developing countries such as India were influential in defining and driving debates on environmental issues at the international level, right from the first global conference on the human environment (UNCHE), held in 1972 in Stockholm. At this summit, then Indian Prime Minister Indira Gandhi delivered a speech that helped shape the dialogue on global environmental issues, including climate change in the subsequent decades. In her speech (Gandhi 1972), by linking environment to poverty and drawing a clear line between the developed and developing countries, the tone was set for future negotiations on most environmental issues, including climate change. This rhetoric dominated the United Nations Conference on Environment and Development (UNCED) summit, held in 1992 in Rio. Importantly, the politics of climate change has primarily been based on disagreements over the ideas that are closely interlinked with issues related to emissions reduction targets, finance, transfer of technology, MRV (measuring, reporting, and verification), and so on. These ideas form the basis of mechanisms such as the Clean Development Mechanism (CDM) and REDD/REDD-plus (reducing emissions from deforestation and forest degradation) among others, which have been opposed and espoused by the emerging economies at various points in time.

The Copenhagen Summit served as a watershed moment in which the BASIC grouping, despite not being a cohesive alliance, emerged in a geopolitical and geoeconomic vacuum, left by the relative decline of the industrialised countries that were reeling under the turmoil created by the 2007–2008 financial crisis. It is necessary first to define what constitutes an emerging economy in the context of this study. The emerging economies have a lower-than-average per capita income, and are largely in transition.[2] They are categorised on the basis of their reforms agenda and rapid growth rates. Moving away from traditional economies, they are on the path to becoming developed economies. However, there is no simple definition of an emerging economy. When the 2007–2008 financial crisis erupted, the emerging economies, especially China and India, were affected only marginally. While China arrived on the world stage as a global economic powerhouse (Barth, Caprio Jr. and Phumiwasana 2009), India's Gross Domestic Product (GDP) grew by 6–8 percent during this period.[3]

The emerging economies, therefore, hinged their climate diplomacy positions on their geopolitical and geoeconomic 'emergence', or 're-emergence' in the case of India and China as these two countries were the largest economies in the world, once upon a time, before colonisation. These countries began to acknowledge their increasing capabilities and, commensurately, growing stature/power in the international political arena and the 'international system' in general, and one of the results of this development was and continues to be the willingness to take on greater responsibility as far as issues of global governance, such as climate change, are concerned. In addition, the emerging economies began to come under increasing pressure to adopt greater commitments to reduce emissions, in recognition of their escalating emissions. At the same time, there has been a greater understanding of the vulnerabilities and risks associated with climate change and an enlargement of pool of scientific knowledge on this issue at the domestic level. Moreover, the increasing role of civil society, sub-national agencies/institutions, and pressure groups (media, business etc.) in steering debates and discussions on climate change has pushed countries such as Brazil and India to improve domestic governance structures related to climate policy. During the period from Copenhagen to Paris and beyond, the emerging economies have played an instrumental role in driving the negotiations agenda – from delineating several facets of 'differentiation' to proposing new mechanisms such as the International Solar Alliance.

Among these countries, India has been in a rather unique position. Its foreign policy agenda has had a distinct influence on its climate diplomacy as well. Ideas of non-alignment (during the Cold War) and revisionism (in the post–Cold War era) are noteworthy in this context. Apart from these geopolitically motivated ideas, it has equally drawn upon civilisational and historical discourses and ideological positions related to the environment (and notions of sustainability) to promote climate diplomacy and create a position for itself in the climate order. Yet, internationally, India has been perceived by several actors as a "spoiler" or "obstructionist". From that position, India's portrayal as a "bridging nation" in the run-up to the Paris Summit is interesting. India's ability to bridge "the many nations across the world" as well as to bridge "development with climate action" (Jayaram 2018, p. 184) came into the limelight. India's climate diplomacy positions are perhaps more influenced by geopolitical and foreign policy imperatives that are replete with ideological viewpoints. For this reason, a case study of India would help better contextualise the ideas that drive the emerging economies' climate diplomacy positions.

Theoretical and methodological framework

A realist-constructivist approach to climate diplomacy

What needs to be highlighted at this juncture is the approach of realist constructivism, whose foundation is built upon two propositions or assumptions. First, constructivism is not a paradigm; it is rather an approach that is compatible with a variety of paradigms, including realism. Second, constructivism need not always be idealist; it is compatible with classical realism that accepts moral ideals as a part of international relations or politics. According to Barkin (2003, p. 337), "realist constructivism would look at the way in which power structures affect patterns of normative change in international relations and, conversely, the way in which a particular set of norms affect power structures". He argues that constructivism emerged in opposition to neorealism or structural realism that has far less space for moral ideals in comparison to classical realism that has in fact many similarities with constructivism. "Political change" cannot be addressed by either "pure realism" or "pure idealism", and therefore, only a combination or, in his words, "interplay of the two" can account for it. In other words, he contends that theories that do not take into account both structures of power and morality become "static" (Barkin 2003, p. 337).

If one takes the case of climate change negotiations, the idea of 'fairness' is linked to the principles of 'historical responsibility' of the developed countries for climate change (in terms of greenhouse gas emissions) and 'relative vulnerability' (physical, socio-economic etc.) of the developing countries to climate change. It is as much about material forces such as the developing countries' right to their share of carbon space for 'development' and 'financial and technological assistance' from the developed countries. As much as 'historical responsibility' is linked with the demand for 'financial and technological assistance' and 'relative vulnerability', with the need for 'development' domestically, these ideational positions, even when they are independent, are considerably shaped by the persisting 'power' equations within the international system. The inherent hierarchical power relationship between the developed and developing countries that has persisted since the Industrial Revolution, during the colonial era and even after the Second World War in which the latter have been subordinate to the former, cannot be discounted in the analysis of 'fairness' (Anderson, Bernauer and Balietti 2017; Paavola and Adger 2006; Parks and Roberts 2006). If 'fairness' is stretched further to analyse the case of emerging powers, the discourse becomes more complicated as their relative growth in power or capabilities distances them further from the developing world (consisting of less and least developed countries), raising the question of 'fairness based on capacity' – whether they should be expected to take on more commitments based on their improved national circumstances (Hochstetler and Milkoreit 2015).

This study further attempts to break the barriers between ideality and materiality that are primarily responsible for casing the relationship between realism and constructivism into a dichotomous one. Barkin (2010, p. 3) calls for a synthesis of realism and constructivism in order to better explain the "relationship between the study of power politics and the study of ideals in international relations on the one hand, and the study of the social construction of international politics on the other". The relative, social, and relational nature of 'power politics' is at the centre of the argument used by Barkin to bridge the differences between realism and constructivism. Since most realists, in general, recognise that context matters to power politics and that mere material assets or capabilities cannot determine outcomes, the relative and relational dimensions are established. Contexts are shaped by ideas and understandings about power – its intrinsic elements, purposes, instruments, and operational modalities. For instance, Chinese scholars developed the concept of "comprehensive national power" or *zonghe gouli* in order to measure the country's national power in 'relation' to that of other major powers and explain the phenomenon of a "rising China" based on their understandings of the concept from a Chinese perspective. By criticising the concept of national power developed by Western scholars, China attempts to build a model that is not confined to material capabilities alone but also includes non-material forces or soft power, including culture (Golden 2011, p. 98).

If ultimately, power politics is about "getting others to do what you want them to do", "persuasion" becomes more critical than "brute force", which can at best be treated as "rhetorical" in many contexts (Barkin 2010, p. 35). With the emergence of 'soft power' as a powerful tool for exercising power in the international system, persuasion or attraction has gained momentum as a useful strategy in relation to force or coercion. Here, Mattern (2005, p. 583) has put forth the idea of "representational force – a nonphysical but nevertheless coercive form of power that is exercised through language" – which could lead to a "reformulation" of soft power as a "continuation" of hard power, rather than as being opposed to it, which is a classic case of realist constructivism. The ability of constructivist approaches to explain and predict could be significantly enhanced by adopting an approach that sufficiently complements the "material–ideational nexus": "the interaction between changing material structures and ideas" (Meyer and Strickmann 2011, p. 67).

In short, realist constructivism facilitates neutralisation of obstacles between ideality and materiality, which is pertinent in the case of climate diplomacy too. It provides an analytical framework for studying the use of power (in its various forms) by nation states to pursue a certain kind of foreign policy (or climate diplomacy as in this case). When it comes to seminal concepts in international relations such as national interest, the approach of realist constructivism is suited to providing an overview of both material and social forces that constitute national interest. National interests, when seen through the constructivist lens, is defined by Weldes (1996, p. 280) as "social constructions created as meaningful objects out of the intersubjective and culturally established meanings with which the world, particularly the international system and the place of the state in it, is understood". However, without accounting for the external reality in materialist or realist terms (defined at times as power politics), the social constructions in the form of discourse or norms cannot stand on their own. In climate diplomacy, national interest is defined in both material (development/Gross Domestic Product, strategic/political autonomy, sovereignty etc.) and moral (climate justice, equity etc.) terms. Another example is that of the formation of 'flexible' alliances in the international climate order based on common or shared 'interests' defined usually in terms of ideas such as common but differentiated responsibilities and respective capabilities (CBDR-RC), equity, and climate justice, among others, whether it is the BASIC grouping (Xinran 2011) or Like-Minded Developing Countries (LMDCs). Historically, the emerging powers – as a part of the G-77 (group of developing countries) – have been known to be opposed to norms, which countries such as India perceived as "one worldism" and "high minded internationalism" (Agarwal and Narain 1991). These norms infringe upon the developing countries' sovereignty, whether it is in the case of imposing legally binding emissions reduction targets on them or measurement, reporting, and verification (MRV) on their domestic actions (Hochstetler 2012). Furthermore, the developing countries pushed for greater equity and justice in climate action through CBDR-RC by urging the developed countries to act more and provide assistance to the developing countries to accelerate poverty reduction programmes among others (Diez 2014). They can be effectively seen as realist constructivist 'constructs' in which ideas and power politics interact with each other. The members of these alliances do not necessarily share cultural or political similarities but share a common agenda of presenting a unified opposition to certain demands of developed or industrialised countries, and thus ensuring survival in the international climate order. Nevertheless, the bottom line here is that both realist views of survival, security and power as well as normative content in the form of morals and ethics are equally important in climate diplomacy. As Barkin (2010, pp. 36–37) observes:

> Politics…requires agency, and agency in turn requires both ideas and materiality. Agency requires ideas because without thought as motivation, there is no agency, there is only inertia and reaction. And it requires materiality because ultimately one cannot dispense with assumptions about human material physiology. Theories of discourse require a material predisposition to communication skills. Theories that look to norms as a key element of political behaviour assume a biologically based sociability that makes normatively based interaction possible. Trying to establish

ontological priority for one or the other creates a chicken-and-egg problem (or, to use a more topical metaphor, an agent-and-structure problem); it is an infinitely recursive process that fails to yield useful insights.

An emerging economy in the international climate order, when seen through the prism of realism, can be conflated with a "rising power" with a "revisionist tendency to alter the status quo" (Lee 2016, p. 103). Schweller (2015, p. 2) calls emerging powers "latecomers" who compete against the "established" powers. He continues:

> Latecomers, whose international prestige lags behind their newly augmented and still improving actual power, typically seek to extend their control and influence over more territory, other states, the world economy, and the set of rules and rights that govern interactions among states (e.g., international norms and regimes, the nature of diplomacy, and property rights on a global scale).

The emerging economies may therefore be expected to change the status quo in the international system or order, even though it is not always the case for various reasons. Their attitudes and behaviour are, however, not just dictated by the constraints of the internal system but also their identities and ideologies commensurate with their relatively growing power or capabilities. On the one hand, they are discontented with their position in the international order and the overall legitimacy of the international order itself, as they 'believe' that their say in decision-making concerning issues of global governance should be higher, commensurate with the relative growth in their material (mainly economic) and power capabilities. On the other hand, they are yet to match the full scope of capabilities that typically are characteristic of industrialised countries. They have to deal with domestic politics and myriad domestic challenges that need to be and are prioritised to a great extent. They are sceptical of the price they might have to pay for enhanced prestige, which incidentally comes with enhanced responsibilities. Finally, they might choose to soften their revisionist tendencies in order to make maximum gains from the existing international order that has facilitated their 'rise' in the first place (Schweller 2015, pp. 5–6). It is rather clear that ideas, perceptions, and attitudes interact with the power capabilities in order to guide state behaviour, and realist constructivism as an approach can thus be used to explain (not necessarily predict) state behaviour in this context.

'Framing' and the methodological framework of the study

Beyond the acknowledgement of climate change as an environmental problem, it has been framed to cover social, economic, moral, political, cultural, ethical, and other dimensions that have substantial implications for policy choices and outcomes through climate diplomacy. The impact of ideas on climate diplomacy is processed through various acts, including 'framing', which is used to influence attitudes and behaviours among decision-makers. Framing is specially a critical part of the agenda-setting process in forums such as the UNFCCC. Climate change itself has been framed in different ways that are reflected in the ideas that govern climate diplomacy. Rein and Schön (2002, p. 146) define framing and frame as follows:

> Framing is a way of selecting, organizing, interpreting, and making sense of a complex reality to provide guideposts for knowing, analysing, persuading, and acting. A frame is a perspective from which an amorphous, ill-defined, problematic situation can be made sense of and acted on.

When contextualised and operationalised in the realm of climate diplomacy, different types of framing assume significance in terms of the choices made by parties to put forth, defend, and legitimise their positions. For instance, equity is framed differently by different emerging economies. Similarly, the continuous framing of climate change as a scientifically uncertain threat or security risk is known to have produced counterproductive results, but its framing as good business or as an opportunity to reduce costs and increase efficiency has led most actors to join the bandwagon of international climate action through financial, technological, and other incentivised mechanisms. When climate change is framed as a sustainable development issue, it is tied with concerns of poverty, equitable distribution of resources (including access), and other socio-economic issues that lie between development and environment.

Pettenger's edited volume on *The Social Construction of Climate Change* (2007) delves into framing of climate change in the industrialised world, and covers the processes, actors, and structures associated with the framing of climate change at the domestic level in countries such as the US, Germany, the Netherlands, and Japan among others. It provides insight into framing and reframing of climate change in terms of sovereignty and territoriality; or neo-colonialism; or ecological modernisation, green governmentality, and civic environmentalism.[4] These frames, based on both external and internal forcings such as human–nature interaction and knowledge creation, are 'explained' and 'uncovered' with the assumption that the understandings of climate change are mainly dominated by the norms and values of the developed countries, especially in the negotiations. Moreover, the changes in norms and values are attributed to processes of social construction (Cass and Pettenger 2007).

This study tries to analyse the changing dynamics of the climate change negotiations, with respect to the frames used by the emerging economies to put forth and legitimise their positions, which gained prominence during the Copenhagen Summit and continued to set the tone of the negotiations since then. Framing in this case is also closely associated with consensus formation: first, among the emerging economies themselves; and second, among the larger international community, consisting of other developing and least developed countries as well as the developed ones. As framing creates different world views, it could be argued that "a frame does not determine a particular position on a substantive policy issue, and many policy positions may be consistent with a given frame" (Rein and Schön 2002, p. 151). Therefore, it is at times difficult to analyse shifts of frame and their impact on policy-making or policy outcomes.

However, this study makes an attempt at tracing policy shifts to changes in frames and framing by altering attitudes and political behaviour. In this context, it is essential to point out that consensus formation through framing is not driven entirely by persuasion; in fact, accommodation has been the purpose of framing in many cases (Charnysh, Lloyd and Simmons 2015), including in the climate change negotiations. For instance, the framing of climate change as an 'opportunity' has won more supporters (in terms of both numbers and diversity) in favour of climate action at the international and national levels than any other frame, as is analysed in this study. The study, therefore, also delves to a minimal extent into the processes of accommodation of different ideas, values, and beliefs, for agenda-setting.

Since the primary objective of the study is to determine the role of ideas in shaping the climate diplomacy positions of the emerging economies and in driving the shifts in their positions in the period leading up to the Copenhagen Summit in 2009 and, more importantly, during 2009–2019, discourse analysis is adopted as a research tool. It helps provide a historical, geopolitical, and socio-economic context to the study, by tracing the evolution of the climate change negotiations. It analyses the process of the coming together of the emerging economies (in the form of BASIC). Discourse analysis is also central to the 'framing' of climate change, as agenda-setting in climate diplomacy is influenced by how climate change is framed by different parties. In this respect, it is essential to bring out differing perspectives on not only the envisaged process(es) and outcome(s) of the climate diplomacy but also how and what meanings are infused into the final outcome(s), in consonance with each nation state's pronounced national interests, defined in terms of a hybrid of ideational and material elements.

This study does not go deep into the nitty-gritty of the technical and legal aspects of the climate change negotiations. Therefore, it delves into the negotiations process, only in terms of balancing of concerns, interests, agreements, and disagreements among the parties, especially in agenda-setting (setting and defining goals as well as action plan and the tools/mechanisms to achieve the goals). More importantly, the study takes a macro perspective to analyse the climate diplomacy positions of the emerging economies, as a part of their larger foreign policy and diplomatic agenda, taking into consideration geopolitical and geoeconomic realities, by equally prioritising the domestic determinants of their climate diplomacy agenda.

While the descriptive part of the study looks into the drivers of and shifts in the climate diplomacy positions of the emerging economies, the analytical part focusses on how ideas have played a bigger role in guiding these positions. It accounts for the shifts in their conduct of politics by analysing the role of ideas that are embedded within and shape the material configurations, and in turn are conditioned by them as well. The approach is therefore, as much interpretive or interpretivist as it is analytical and descriptive, as it reflects on the purposes and processes of 'framing' issues and ideas in climate diplomacy, based on differing perspectives. The qualitative approach complements the fundamental and deductive type of research used in it, which implies discourse analysis of shifts based on empirical data and indicators that connect the observable phenomena with theory. Herein, the establishment of centrality of ideational factors, which are not disconnected from material forces, to the shifts is accomplished in tandem with the evidenciation and explication of these shifts.

The data used in the study are gathered from both primary and secondary sources. A large proportion of data is collected through interviews, policy documents (such as agreements or action plans), and official statements (attributable to government representatives/ agencies). At the same time, peer-reviewed journals and reports/policy briefs etc. are also utilised. By conducting in-depth interviews – mostly semi-structured – with academics/researchers, policy-makers and bureaucrats, and non-governmental activists, the discourse on the emerging economies' climate diplomacy positions is assimilated into the study and analysed through qualitative methods. The qualitative nature of data collection employed in this study is characterised by open-ended questions, which also helps provide an overview of variations in interpretations of similar actions. By using non-probability sampling, particularly expert, purposive, and snowball sampling, the interviewees have been chosen based on their experience and expertise in the climate change negotiations and climate diplomacy in general. Interviews have been conducted with practitioners and researchers from not only the BASIC countries but also other countries, especially in the European Union.

An in-depth 'case study' (research design) of India has been used to obtain result(s), based on which a generalisable pattern could be generated that could be applied to other emerging economies' climate diplomacy positions as well. The study uses categorisation and summarisation of the data to identify common themes and meanings/interpretations of ideas that emerge in the data collected from various sources, as well as to describe and validate the observable phenomena. The summarisation also facilitates recognition of "formula" in discourse analysis. Krieg-Planque's notion of "formula" – as translated by Oliveira (2018, p. 47) – provides a theoretical and methodological framework to a study, and is defined as "a set of formulations used in a time and in a specific public space, crystallising political and social issues". A formula, when used in this study, can be treated as frames or formula narratives (into which the above-mentioned ideas are circumscribed), shaped or influenced by the contexts, which provide the basis for attitudes, behaviours, and positions of the emerging economies in climate diplomacy. By using the theoretical framework of realist constructivism and discourse analysis, the functioning of the formula in climate diplomacy (and the role of ideas in it) can be described, as the formula essentially condenses discourses on the ideas in focus. Furthermore, as also acknowledged by Oliveira, a work based on this method, on the notion of discursive formula, "can straddle the enunciative and materialist approaches" (2018, p. 51).

To provide a background to analysis of 'formula narratives' using discourse analytical approach, a pattern with regard to the positioning of the emerging economies is elaborated upon here. Ideas such as equity, climate justice, sovereignty, solidarity, leadership, and vulnerability, and their perceptions are embedded in each formula that is prevalent in climate diplomacy as a practice, not as fixed discourses but as living, dynamic, and consequential discursive forces. Within a formula, recurring patterns of interpretation, conflicting perspectives, and power relations are intrinsically entrenched, thereby allowing the researcher to delve into both the 'what' and 'how' of the analysis. These formula narratives are constituents of the discourses that guide, organise, and legitimise climate diplomacy positions adopted by nation states (emerging economies in this study).

The study therefore focuses on the shifts or changes in the 'formula narratives' used by the emerging economies to position themselves in the international climate order during the period identified in it. As frame analysis is central to this study, it has contextualised and analysed these narratives (based on the five categories of ideas) within the scope of three objectives: diagnostic, prognostic, and motivational framing (Benford and Snow 2000). Under diagnostic framing, perceptions about climate change, the causes for it, and so on are looked at. Under prognostic framing, perceptions about solutions, tactics, and strategies to address climate change are dealt with. And under motivational framing, a call for climate action and the rationale for it are under enquiry (as there is an agreement on the diagnostic and prognostic frame). For instance, the emerging economies, under the diagnostic framing, have framed climate change in different contexts, as a development issue (socio-economic idea); as a geopolitical issue, blaming the developed world for their contribution to the problem and preventing the developing countries from developing (geopolitical idea); or as an environmental (vulnerability) problem (ecological/environmental idea) etc.

The shifts in climate diplomacy positions of the emerging economies also reflect shifts in these frames. For instance, from an issue on which India felt it was not obliged to act, climate change became the centre-point of framing leadership through promises and actions such as an emissions intensity (of its GDP) reduction target in 2009 and the establishment of the International Solar Alliance in 2015, in which India played the primary role. Temporally, the study mainly focusses on the period between the Copenhagen Summit (2009) and Madrid Summit (2019), with references to the period between 1992 and 2009, where deemed necessary. For instance, the analysis of the shifts in the climate diplomacy positions of the emerging economies has to account for the Bali Action Plan (2007) in which the meaning of differentiation between the developed and developing countries was changed for the first time, or the 2007–2008 financial crisis that had an impact on the global geoeconomic order. It allows a multilevel analysis of the conduct of climate diplomacy, thereby also providing a policy-oriented framework of climate action, carried out through diplomatic channels and agencies, without constraining the understanding of this concept only to multilateral climate change negotiations.

As far as the case study of India is concerned, the study goes beyond the international dimensions of its climate diplomacy, to look at its bilateral and other multilateral engagements in this domain, but to a lesser extent. In addition to India's relations with other members of the BASIC countries, it deals with India's climate diplomacy with the US and the EU. There are two reasons why the study goes beyond the realm of the international climate change negotiations. First, climate diplomacy at other levels is not only growing day by day (in terms of the agreements signed) but also becoming more and more influential, even in terms of the multilateral efforts. This is evidenced by the greater attention being paid to climate change in the foreign policies of the majority of nation states and its translation into cooperation in various sectors pertaining to climate change mitigation and adaptation. Second, the interface between the negotiations and other forms of climate diplomacy is relevant for a better understanding of the positions taken by nation states at the former. For instance, the agenda-setting process for the international climate change negotiations sometimes begins in other forums – a group of like-minded countries sharing common concerns, or regional groupings developing a common position, or major countries seeking to arrive at a consensus jointly. By doing this, the impact of multilevel diplomatic engagements on the shifts in the positions taken by the emerging economies in the negotiations has been analysed in the study. In essence, the study meanders through the conceptual and practical panorama of climate diplomacy of the emerging economies; and the drivers of and shifts in their positions in this domain.

Outline of the study

The study has a three-tiered framework. First, it analyses "climate diplomacy" of the emerging economies, using Barkin's (2010) alternative to "paradigmatic way of thinking about different approaches to the study of international relations" that provides a framework to analyse climate diplomacy that combines the elements of both realism and constructivism. Second, the study seeks to analyse the role of the emerging economies in the international climate order empirically. Under this, the BASIC countries are studied in-depth, both collectively and individually. These countries' positions from the Copenhagen Summit to Paris Summit and beyond, explanations and drivers of their positions (both domestic and international), as well as convergences and divergences among them are explored in order to gauge the contributory role of ideas and materiality in shaping their climate diplomacy agenda.

Third, the study uses the case study of India to elaborate further upon the research question of whether ideas are more critical to defining climate diplomacy positions of emerging economies than material interests. It deals with the questions of whether ideas are influenced by power politics or vice versa as far as climate diplomacy is concerned; and how ideas and material interests interact with each other in their practice of climate diplomacy. The study charts the shifts in India's climate diplomacy positions during 2009–2019, by focussing not only on the climate change negotiations but also bilateral dynamics that India shared with other major players during this period. The study concludes that ideational forces have been more influencing in shaping the emerging economies' climate diplomacy agenda. The chapterisation of this study has been designed broadly on the basis of the research objectives and research design identified and selected, respectively – focussing on the emergence of the BASIC countries in the international climate order, and a country-wise analysis. This is followed by an in-depth analysis of the Indian case study in the following two chapters – divided into two periods (2009–2013 and 2014–2019).

Notes

1 More information regarding China–EU cooperation on climate change can be found on the European Commission's website: https://ec.europa.eu/clima/policies/international/cooperation/china_en.

2 Many would argue that China should no longer be considered an emerging economy due to its "steady economic growth", its status of "upper middle-income economy", and so on (see Ahlstrom 2020). However, for most of the period (2009–2019) chosen for this study, China's status was largely acknowledged as that of an emerging economy. Even today, although China's GDP per capita has passed USD 10,000, it continues to remain a middle-income country according to the World Bank.

3 India's GDP growth (annual percentage) can be found on the World Bank's website: https://data.worldbank.org/indicator/NY.GDP.MKTP.KD.ZG?locations=IN.

4 The discourse on ecological modernisation emphasises the "compatibility between economic growth and environmental protection, or more specifically between a liberal market order and sustainable development". It is based on the premise that "environmental degradation can be decoupled from economic growth, and that capitalism and industrialization can be made more environmentally friendly through green regulation, investment and trade". Green governmentality presents a case of "a global

form of power tied to the modern administrative state, mega-science and the business community", exemplified by a multilateral climate regime. Civic environmentalism embodies the radical and reform-oriented narratives that are sceptical of both green governmentality and ecological modernisation. They call for a redefinition of "the basic principles of climate governance towards equity and ecological sustainability", thus reframing the scope and contours of contemporary environmental practices (Bäckstrand and Lövbrand 2007).

Bibliography

adelphi. n.d. "About Climate Diplomacy.". *Climate Diplomacy*. Accessed December 20, 2019. https://www.climate-diplomacy.org/about-climate-diplomacy.

Agarwal, Anil and Narain, Sunita. 1991. "Global Warming in an Unequal World: A Case for Environmental Colonialism." *Centre of Science and the Environment*. January 1. Accessed June 20, 2020. http://www.indiaenvironmentportal.org.in/content/253824/global-warming-in-an-unequal-world-a-case-of-environmental-colonialism/.

Ahlstrom, David. 2020. "Why China Is No Longer an Emerging Economy." *CUHK Business School*. March 5. Accessed August 5, 2020. https://cbk.bschool.cuhk.edu.hk/why-china-is-no-longer-an-emerging-economy/.

Anderson, Brilé, Bernauer, Thomas and Balietti, Stefano. 2017. "Effects of Fairness Principles on Willingness to Pay for Climate Change Mitigation." *Climatic Change* 142 (3): 447–461.

Bäckstrand, Karin and Lövbrand, Eva. 2007. "Climate Governance Beyond 2012. Competing Discourses of Green Governmentality, Ecological Modernization and Civic Environmentalism." In *The Social Construction of Climate Change: Power, Knowledge, Norms, Discourses*, edited by Mary E. Pettenger, 123–147. Hampshire: Ashgate Publishing Ltd.

Barkin, Samuel J. 2003. "Realist Constructivism." *International Studies Review* 5 (3): 325–342.

Barkin, Samuel J. 2010. *Realist Constructivism: Rethinking International Relations Theory*. Cambridge: Cambridge University Press.

Barth, James R., Caprio Jr., Gerard and Phumiwasana, Triphon. 2009. "The Transformation of China from an Emerging Economy to a Global Powerhouse". In *China's Emerging Financial Markets: Challenges and Opportunities*, edited by James R. Barth, John A. Tatom and Glenn Yago, 73–110. Berlin: Springer.

Belis, David, Joffe, Paul, Kerremans, Bart and Qi, Ye. 2015. "China, the United States and the European Union: Multiple Bilateralism and Prospects for a New Climate Change Diplomacy." *Carbon & Climate Law Review* 9 (3): 203–218.

Belis, David, Shunz, Simon, Wang, Tao and Jayaram, Dhanasree. 2018. "Climate Diplomacy and the Rise of 'Multiple Bilateralism' between China, India and the EU." *Carbon & Climate Law Review* (Lexxion) 12 (2): 85–97.

Benford, Robert D., and Snow, David A. 2000. "Framing Processes and Social Movements: An Overview and Assessment." *Annual Review of Sociology* 26: 611–639.

Bush, Richard C. 2011. "The United States and China: A G-2 in the Making?" *Brookings*. October 11. Accessed August 2, 2020. https://www.brookings.edu/articles/the-united-states-and-china-a-g-2-in-the-making/.

Cass, Loren R., and Pettenger, Mary E.2007. "Conclusion: The Constructions of Climate Change." In *The Social Construction of Climate Change: Power, Knowledge, Norms, Discourses*, edited by Mary E. Pettenger, 235–246. Hampshire: Ashgate.

Charnysh, Volha, Lloyd, Paulette and Simmons, Beth A. 2015. "Frames and Consensus Formation in International Relations: The Case of Trafficking in Persons." *European Journal of Intrnational Relations* 21 (2): 323–351.

Diez, Cristina. 2014. "Policy Brief and Proposals: Common but Differentiated Responsibilities." *ATD Fourth World*. March 13. Accessed August 24, 2020. https://sustainabledevelopment.un.org/getWSDoc.php?id=4086.

Elliott, Lorraine. 2013. "Climate Diplomacy." In *The Oxford Handbook of Modern Diplomacy*, edited by Andrew F. Cooper, Jorge Heine and Ramesh Thakur, 840–856. Oxford: Oxford University Press.

Gandhi, Indira. 1972. "Man and Environment." *LASU-LAWS Environmental Blog*. June. Accessed December 20, 2018. http://lasulawsenvironmental.blogspot.com/2012/07/indira-gandhis-speech-at-stockholm.html.

Golden, Sean. 2011. "China's Perception of Risk and the Concept of Comprehensive National Power." *The Copenhagen Journal of Asian Studies* 29 (2): 79–109.

Hochstetler, Kathryn A. 2012. "The G-77, BASIC, and Global Climate Governance: A New Era in Multilateral Environmental Negotiations." *Revista Brasileira de Política Internacional* 21 (2): 53–69.

Hochstetler, Kathryn A. and Milkoreit, Manjana. 2015. "Responsibilities in Transition: Emerging Powers in the Climate Change Negotiations." *Global Governance* 21 (2): 205–226.

Hsu, Angel, Moffat, Andrew S., Weinfurter, Amy J. and Schwartz, Jason D. 2015. "Towards a New Climate Diplomacy". *Nature Climate Change* 5 (6): 501–503.

Hurrell, Andrew and Sengupta, Sandeep. 2012. "Emerging Powers, North-South Relations and Global Climate Politics." *International Affairs* 88 (3): 463–484.

Jayaram, Dhanasree. 2018. "From "Spoiler" to "Bridging Nation": The Reshaping of India's Climate Diplomacy." *La Revue Internationale et Stratégique* 1 (109): 181–190.

Jayaram, Dhanasree. Forthcoming. "India's Climate Diplomacy towards the EU: From Copenhagen to Paris and Beyond." In *EU-India Relations: The Strategic Partnership in the Light of the European Union Global Strategy*, edited by Philipp Gieg, Timo Lowinger, Manuel Pietzko, Anja Zürn, Ummu Salma Bava and Gisela Müller-Brandeck-Bocquet. Cham: Springer.

Karakır, İrem Aşkar. 2018. "Environmental Foreign Policy as a Soft Power Instrument: Cases of China and India." *Journal of Contemporary Eastern Asia* 17 (1): 5–26.

Karlsson, Christer, Parker, Charles, Hjerpe, Mattias and Linnér, Björn-Ola. 2011. Looking for Leaders: Perceptions of Climate Change Leadership among Climate Change Negotiation Participants. *Global Environmental Politics* 11 (1): 89–107.

Kuyper, Jonathan W., Linnér, Björn-Ola and Schroe, Heike. 2017. "Non-state Actors in Hybrid Global Climate Governance: Justice, Legitimacy, and Effectiveness in a Post-Paris era." *WIREs: Climate Change* 9 (1): e497.

Lee, Paul S. N. 2016. "The Rise of China and Its Contest for Discursive Power." *Global Media and China* 1 (1–2): 102–120.

Mabey, Nick, Gallagher, Liz and Born, Camilla. 2013. *Understanding Climate Diplomacy: Building Diplomatic Capacity and Systems to Avoid Dangerous Climate Change.* London: Third Generation Environmentalism.

Mattern, Janice Bially. 2005. "Why 'Soft Power' Isn't So Soft: Representational Force and the Sociolinguistic Construction of Attraction in World Politics." *Millennium: Journal of International Studies* 33 (3): 583–612.

Meyer, Christoph O. and Strickmann, Eva. 2011. "Solidifying Constructivism: How Materialand Ideational Factors Interact inEuropean Defence." *Journal of Common Market Studies* 49 (1): 61–81.

Oliveira, Helio. 2018. "The Black Consciousness Movement in Brazil: A Materialist-Enunciative Approach to Discourse Analysis." In *Material Discourse – Materialist Analysis: Approaches in Discourse Studies*, edited by Johannes Beetz and Veit Schwab, 47–64. Lanham: Lexington Books.

Paavola, Jouni and Adger, W. Neal. 2006. "Fair Adaptation to Climate Change." *Ecological Economics* 56 (4): 594–609.

Parks, Bradley C. and Roberts, J. Timmons. 2006. "Globalization, Vulnerability to Climate Change, and Perceived Injustice." *Society & Natural Resources* 19 (4): 337–355.

Pettenger, Mary E. 2007. "Introduction: Power, Knowledge and the Social Construction of Climate Change." In *The Social Construction of Climate Change: Power Knowledge, Norms, Discourses*, edited by Mary E. Pettenger, 1–22. Hampshire: Ashgate.

Rein, Martin and Schön, Donald. 2002. "Reframing Policy Discourse." In *The Argumentative Turn in Policy Analysis and Planning*, edited by Frank Fischer and John Forester, 145–166. London: UCL Press.

Renner, Michael. 2015. "Connecting the Dots: Integrating Green Jobs into Climate Diplomacy." *Climate Diplomacy.* February 11. Accessed August 4, 2020. https://www.climate-diplomacy.org/news/connecting-dots-integrating-green-jobs-climate-diplomacy.

Schweller, Randall L. 2015. "Rising Powers and Revisionism in Emerging International Orders." *Valdai Papers* 16. May 20. Accessed August 4, 2020. https://valdaiclub.com/a/valdai-papers/valdai_paper_16_rising_powers_and_revisionism_in_emerging_international_orders/.

Weldes, Jutta. 1996. "Constructing National Interests." *European Journal of International Relations* 2 (3): 275–318.

Xinran, Qi. 2011. "The Rise of BASIC in UN Climate Change Negotiations." *South African Journal of International Affairs* 18 (3): 295–318.

Zhang, Hai-Bin, Dai, Han-Cheng, Lai, Hua-Xia, Wang, Wen-Tao. 2017. "U.S. Withdrawal from the Paris Agreement: Reasons, Impacts, and China's Response." *Advances in Climate Change Research* 8 (4): 220–225.

2 Emerging economies in the international climate order

The emergence of BASIC countries in the international climate order

Before going into the drivers of BASIC's formation and its evolution during 2009–2019, a brief description of this grouping and its rise during the Copenhagen Summit as the most decisive group would provide the basis for why it is important to study them in the context of this study. The shifts in climate diplomacy positions of the emerging economies can also be best analysed when seen from a historical perspective, especially since the Bali Action Plan (2007). The Copenhagen Summit was seen as symbolic of the emergence of a new political order, "one shaped by the new self-confidence of the Asians and the powerlessness of the West".[1] Power politics took centre-stage, as indicated by the emerging economies' show of strength to push its agenda on 'climate justice' and 'equity', which were perceived to be under threat. It must be noted that the leaked Danish text (The Guardian 2009) pointed towards a move to grant the responsibility of the administration of climate finance to Bretton Woods institutions such as the World Bank and/or International Monetary Fund (IMF). These international financial institutions continue to be governed by the financial contribution of each country and thus are considered to be biased towards the 'rich' countries. Countries such as India and China have questioned these institutions' monopoly repeatedly. They have also championed the cause of a New International Economic Order (NIEO) by revising the existing financial/economic order, an idea that took birth in the 1970s – calling for a more equitable distribution of resources in the world, with a focus on bridging the widening gap between the developed and developing countries (UN 1974). In the climate change arena, even before the Green Climate Fund was formed in 2010, the emerging economies, as a part of the G-77 group of developing countries, have expressed scepticism over the legitimacy of the Global Environmental Facility (GEF), which was established in 1992 to mobilise grants and finance for sustainable development initiatives (Gupta 1997). The developed countries can exert influence on the developing countries' economic policies through their development assistance, which could even jeopardise socio-economic advancement for environmental gains.

The emerging economies have their competing interests and geopolitical divergences (such as the Sino-Indian border dispute). However, their long-term cooperation, especially in the form of South–South cooperation (Cui 2016), contributed to their ability to coordinate and project a common stance at the Copenhagen Summit. The majority of countries in G-77 are postcolonial societies that at times share deep-seated mistrust of the West, more so of their European colonisers. Having emerged from colonialism, which had a negative impact on their economies and societies, they viewed any attempts to impose environmental agreements on them as a deliberate attempt to prevent them from developing so that they would never be in a position to challenge the position of the industrialised countries in the international system. The fact that the colonial era in these countries also saw large-scale damage to their ecosystems and environment – perpetrated by the colonisers – is also held against the industrialised countries, many of which were colonisers (Gupta 1997, p. 183). However, more recently, it is the quest for sharing of knowledge, technology, and resources among the developing countries, as well as promotion of trade and commerce, that brought the emerging economies on to a common platform. A number of groupings such as IBSA (India, Brazil, and South Africa) and BRIC (Brazil, Russia, India, and China; now BRICS after the addition of South Africa) emerged in the 2000s to carry forward the spirit of South–South cooperation, as a corollary to North–South divide, which has proven to be difficult to bridge (Gray and Gills 2016).

Having established cooperation and coordination through the IBSA Dialogue Forum in agriculture, trade, technology, culture, and defence among others, India, Brazil, and South Africa had already managed to nurture mutual trust amongst themselves (Mishra 2018). According to the official website of the forum, it is rooted in the identity of these countries as "developing, pluralistic, multi-cultural, multi-ethnic, multi-lingual and multi-religious nations" (IBSA 2003). IBSA's stated commitment to sustainable development that is equitable, just, and inclusive, extends beyond its sovereign spaces to include other developing countries as well, thereby attempting to bolster 'solidarity' with the developing world by supporting development projects in them. These countries, understandably having attained more resources (especially in the economic sphere), began to visualise the possibility of result-oriented and tangible cooperation among the developing countries with more confidence.

China, with its rapid economic growth and growing economic tentacles all over the world, had begun to make inroads into several countries, including Brazil and South Africa. The G-77 has also served as a platform for China to build financial and institutional linkages with its members since 1994. China's strategy of providing foreign aid to developing countries "with no strings attached" began to challenge the traditional paradigm of foreign aid followed by the Organisation for Economic Cooperation and Development (OECD) countries as well as the Bretton Woods institutions (Kjøllesdal and Welle-Strand 2010). Unlike the aid provided by the latter, China does not impose any conditions on the aid recipients in terms of values such as democracy, transparency, human rights, and 'good governance', thereby ensuring 'principled' non-intervention in their domestic affairs (Zheng 2016).

While this can be seen as China's attempts to increase its influence in the international system through geoeconomic power projection, it is simultaneously a reflection of the principles identified in the final communiqué of the Asian-African Conference in Bandung, Indonesia in 1955, laying the foundation for the Non-Aligned Movement. Among the principles are "respect for the sovereignty and territorial integrity of all nations" and "abstention from intervention or interference in the internal affairs of another country" (Republic of Indonesia 1955). The majority of the countries that participated in this conference were newly independent states that were determined to condemn and oppose intervention and interference at all levels. This can also be considered an exemplification of South–South cooperation, wherein postcolonial states challenge the dominant norms designed by the West. Undoubtedly, this has been made possible by the fast-growing economies of these countries, especially China and India.

Therefore, the emergence of BASIC in the international climate order was not sudden. One of the significant moments in the history of climate change negotiations was the 2007 Bali Summit, wherein for the first time, the developing countries were also called upon to undertake "nationally appropriate mitigation actions...in the context of sustainable development, supported and enabled by technology, financing and capacity-building, in a measurable, reportable and verifiable manner" (UNFCCC 2008, p. 3). In addition, on the decision of "long-term cooperative action"[2] (UNFCCC n.d.), the strict differentiation between developed and developing countries, as in the Kyoto Protocol (KP), was also watered down. The international climate regime was moving towards an architecture that would differentiate between the developing countries and not view them as a single, homogenous group. Hence, developing countries that are economically better off would be expected to commit to emissions reduction in a more proactive manner (perhaps even legally binding); and by linking all forms of support with mitigation efforts, it was made sure that the post-Kyoto regime would put emphasis on making the former contingent on the latter. On the one hand, the ideas of equity and justice were being challenged, especially since proposed approaches based on "equal per capita entitlements"[3] for a post-Kyoto arrangement did not take off. And on the other, 'non-intervention in domestic affairs' and 'respect of sovereignty' were in contention, with changes in positions on transparency and scrutiny of domestic actions brought out by the Bali Action Plan.[4]

On the issue of adaptation, there was no clarity on what would be the future of adaptation finance. Until then, the negotiations were mostly focussed upon mitigation and even the developing countries feared that too much engagement with adaptation could deviate the developed countries' attention from mitigation, which was more important to prevent climate change. Many countries also scuttled the adaptation negotiations track due to "perceived self-interest". The developed countries did so in order to avoid acknowledgement of "responsibility and liability based on historical emissions", which might snowball into demands for resolving other global challenges such as poverty. At the same time, fossil-fuel exporting countries such as the ones in the Organization of the Petroleum Exporting Countries (OPEC) have been demanding "compensation" for diversification of their "oil-dependent economies" as adaptation. In effect, under the Bali Action Plan, through its increased emphasis on adaptation, the negotiations, mainly driven by the developed countries, managed to drive a wedge between the developing countries – by prioritising those countries that are considered most vulnerable to climate change, particularly the Least Developed Countries (LDCs) (Ciplet, Roberts and Khan 2013).

What should also be taken into consideration is the financial crisis that hit mostly the rich countries in 2007. The fallouts of the 2007–2008 financial crisis seem to have a lasting impact on the trajectory of the climate change negotiations. The recession reduced economic activity in the developed world significantly for a short period. During this period, particularly in 2008, the developed countries' growth in emissions decreased, with the developing countries' emissions accounting for more than half of the world's greenhouse gas (GHG) emissions, according to data published by the Netherlands Environmental Assessment Agency (NEAA 2008). The overall impact on the emissions scenario showed positive signs as the growth in global emissions came down by half in this year. It is difficult to ascertain whether the financial crisis alone contributed to the lowering of emissions, as green restructuring of the economies – especially in terms of deployment of renewable energy and enhancement of energy efficiency measures – could have also had an impact, albeit marginal (Clark 2009).

Nevertheless, the growing proportion of developing countries' GHG emissions in the run-up to the Copenhagen Summit is considered to have influenced the discourse on the responsibility of the emerging economies towards finding solutions. China overtook the US as the world's largest emitter of carbon dioxide in 2007 (Jones 2007). India became the third-largest emitter of GHGs in 2009, surpassing Russia, whose emissions reduced due to economic recession (Chauhan 2010). Brazil and South Africa's GHG emissions were also on the rise due to deforestation and excessive coal use, respectively. This put more pressure than ever before on the BASIC countries to act. Even some of the other developing countries had started to become critical of these countries, particular China and India, which did not have any legally binding commitments under the KP. For instance, India has been accused by its South Asian neighbours of holding a "rigid position" at the negotiations and sabotaging the climate agreement. India began to be considered indifferent to the dangers they face due to rising sea levels and so on, scuttling their access to climate funds due to the stalemate (The Third Pole 2015) (Table 2.1).

In addition to the growing emissions of the BASIC countries, their contribution to global Gross Domestic product (GDP) and collective share in global trade grew significantly in the 2000s, partially also due to the economic downturn in the developed economies. They have

Table 2.1 BASIC Countries' GHG Emissions from 1990 to 2010 (including land use change and forestry)

Country	Absolute Change (in tons CO_2e)	Relative Change (in percent)
Brazil	+44.05 million	+3
China	+6.89 billion	+237
India	+1.27 billion	+108
South Africa	+174.08 million	+55

Source: Our World in Data and CAIT Climate Data Explorer.

been among the fastest growing economies in the world, particularly China and India (Hallding et al. 2011, p. 31), leading to the assumption/understanding that with growing emissions and economies, they have a bigger responsibility towards international climate action than before. The idea of emerging powers was emboldened by their material superiority and rising stature in the international system, in comparison to the other developing countries. It was equally strengthened by their engagement with the 'major powers' in forums such as G-20 and Major Economies Forum on Energy and Climate Change (MEF), which placed them separately from the LDCs and Climate Vulnerable Forum (CVF) countries, whose requirements and identities are shaped by varying circumstances.[5]

In 2005, the BASIC countries alone (apart from Mexico) were invited to the G-5 summit by the UK, which later translated into the formation and institutionalisation of the G8+5 Climate Change Dialogue. The logic behind building this dialogue process was that "developing countries are expected to contribute 39 percent of global emissions by 2010, and independent estimates are that in the next 24 years 50 percent of global energy investment will be in developing countries" (Jaura 2006). Therefore, steps towards identifying solutions to GHG emissions through a model of 'shared responsibility' and encouraging low-carbon investments in developing and emerging economies were set in motion. Another development was the initiation of Major Economies Meetings on Energy Security and Climate Change (MEM) under the leadership of former US President George W. Bush in 2007, which was later renamed as MEF by former US President Barack Obama, in which the BASIC countries have a major role to play (Obama 2009). Furthermore, the BASIC countries are active members of G-20. Although this grouping's inception dates back to 1999 (after the Asian financial crisis), its first summit was held in 2008; and since then, the members of G-20 have been meeting regularly and discussing climate change among other issues. The G-20 countries account for almost 90 percent of global GDP, 80 percent of international trade, two-third of the world's population, and 84 percent of all fossil fuel emissions (G20 Foundation n.d.). Among the various realities and changes that the establishment of G-20 recognised was the "growing importance of emerging economies" (Eurostat n.d.).

In addition, the framing of climate change as a security threat or a "threat multiplier" began to gain momentum, with the introduction of the issue in the UN Security Council (UNSC) in 2007, under the United Kingdom's presidency. This move was endorsed by the developed world as well as many developing countries (especially

LDCs). However, the emerging economies such as India and China have largely been opposed to the idea of introducing the security implications of climate change in the UNSC on the grounds that a security approach may endanger climate justice (particularly the principle of CBDR-RC), collective efforts from the global community, development-related concerns of the developing countries, and so on. It was also observed by many representatives at the meeting that the UNSC was and is not the appropriate forum to discuss climate change due to its lack of representation (compounded by the veto powers vested in five countries) (Scott 2018; UNSC 2007). The timing of this meeting raised several questions about its intentions and objectives. By securitising the issue (at the UN), the developed countries – backed by many 'vulnerable' countries – seemed to have not only raised the urgency of the issue, calling for more extreme measures to deal with climate change, but also created greater divisions within the developing world.[6]

In a reaction to these developments, the BASIC countries, particularly China and India felt the need to unite and shift their climate diplomacy positions. From being completely anchored in the G-77 (which was their strategy for a long time to avoid any commitments within a legal framework), they began to put more energy into forming a group outside the G-77 to protect the interests of the emerging economies (Hallding et al. 2011). Hence, the BASIC countries are known to have coordinated their positions in the run-up to and during the 2009 Copenhagen Summit in a highly systematic manner. According to the statement of the former Indian Union Minister of State (Independent Charge), Ministry of Environment and Forests (MoEF), Jairam Ramesh, in the Rajya Sabha (the Indian Parliament's Upper House) on the Copenhagen Accord (The Hindu 2009):

> A notable feature of this Conference that has been widely commented on is the manner in which the BASIC group of countries coordinated their positions. Ministers of the BASIC group comprising Brazil, South Africa, India and China has met in Beijing on November 27th and 28th, 2009 to prepare for Copenhagen in a joint manner....The BASIC Group Ministers met virtually on an hourly basis right through the Copenhagen Conference. Within BASIC, India and China worked very, very closely together. I believe that the BASIC group has emerged as a powerful force in climate change negotiations and India should have every reason to be satisfied it has played in catalyzing the emergence of this new quartet. Their unity was instrumental in ensuring that the

Accord was finalized in accordance with the negotiating framework as laid out in the UNFCCC, Bali Action Plan and the Kyoto Protocol.

In fact, India is regarded as the chief architect of this grouping as it joined hands with China and convinced its IBSA partners – Brazil and South Africa – to agree on disagreeing with the demands of the industrialised bloc (Paskal 2010). These demands included adoption of a legally binding emissions reduction target from the emerging countries based on 'shared responsibility', global emissions peaking date, and 'international' Measuring, Reporting and Verification (MRV) for domestic actions among others. The four emerging economies jointly called on the developed countries to enforce the second commitment period of the 1997 Kyoto Protocol. They insisted that all their mitigation actions would be voluntary and not subject to "intrusive" international scrutiny (Dasgupta 2009).

The US and EU's attempts to divert the debate away from the strict dichotomy between Annex I and non–Annex I parties, and in effect not adopt stronger commitments, were opposed. Neither the US nor the EU foresaw this negotiating move by the BASIC countries that was then steered by the Chinese delegation towards its final execution. In fact, the Obama administration's then Secretary of State Hillary Clinton later admitted in her memoir that the meeting of the BASIC leaders was literally gatecrashed by Obama, a step that ultimately brought the US too on board. In the end, although it was a collective failure, China and India were blamed for "sabotaging" the negotiations (Rapp, Schwägerl and Traufetter 2010). At the same time, it must be noted that the BASIC countries, for the first time, pledged to reduce their emissions or emissions intensity, at the climate change negotiations in 2009. This itself was a change in position for these countries that had long been reluctant to adopt any kind of commitments, domestically or internationally (Table 2.2).

In essence, the COP-15 was a geopolitically critical summit in many ways. India and China have taken similar stands on issues such as CBDR, climate finance, and transfer of technology (ToT). At the Copenhagen Summit, they also helped draft the full text of the Copenhagen accord (albeit with no legal subtexts) together, which would guide the progress of climate change negotiations in the coming years. This was the first time that countries such as China, India, Brazil, and South Africa had broken ranks with the G-77 (ET 2009). Only a few countries from the G-77 bloc had backed the Copenhagen Accord thrashed out by the BASIC countries (The Hindu 2009). The Copenhagen Summit

Table 2.2 BASIC Countries' Emissions Reduction Pledges at the 2009
Copenhagen Summit

Country	Pledge
Brazil	Reduce national emissions by between 36 and 39 percent below 'business as usual' levels by 2020
China	Reduce the 'emissions intensity' of GDP to between 40 and 45 percent below 2005 levels by 2020
India	Reduce the 'emissions intensity' of GDP by 20–25 percent against 2005 levels by 2020
South Africa	Reduce national emissions to 34 percent below 'business as usual' levels by 2020

Source: Climate Action Network.

also gave rise to debates about the contested concept of G-2 that refers to the US and China being at the forefront of the international order although some analysts have also interpreted G-2 as the emergence of China and India (Paskal 2010). Furthermore, the EU was forced to take the backseat during the climate change negotiations. Eventually, it had to be satisfied with the lowest hanging fruit despite coming into the negotiations with high expectations, particularly with respect to emissions cut commitments from countries such as China and, to a lesser extent, India. The European contingent (particularly Western European countries) have had the distinction of leading the negotiations process from the front since early 1990s (Hewitt 2009).

Explaining emerging economies' climate diplomacy positions up until the Copenhagen Summit

There have been attempts to provide an interest-based model to explain the positions of developing countries in the climate change negotiations. According to Rong (2010), both domestic and international factors are critical to their positions. Listing out five hypotheses (p. 4584), Rong concludes that more than ecological vulnerability, it is mitigation capability that influences the developing countries' climate diplomacy positions. According to her hypotheses, factors such as ecological vulnerability, capacity, financial and technological support, growing emissions, and international pressure feature prominently, as the ones influencing developing countries' decisions and pledges on climate action. However, they fail to take into consideration framing of climate change issues that play a crucial role in shaping these factors. For instance, India's growing emissions are not the same as that of China since the gap between the two countries' emissions, especially per capita emissions, is gigantic. This was instrumental in deciding the next steps in strategic climate action taken by these countries in the run-up to the Paris Agreement. While China, in 2014, announced that it would strive to reach peak carbon emissions in or around 2030 (The White House 2014), India asserted that it had no intention to declare the year or estimate the period of peaking emissions as of yet (explained in the subsequent chapters).

A particular country's vulnerabilities and capacities are also influenced by domestic actors that support a set of climate actions in comparison to others. In Brazil, for instance, the share of renewable and clean energy has always been high, in comparison to other emerging economies, primarily fed by hydropower and biofuels. The major concern for the country in the climate change arena has been deforestation in the Amazon rainforests, one of the ecological hotspots of the world. However, since 2005, reasons including steady economic growth, "declining prices of export commodities such as soy and beef", and a series of domestic legislative actions taken on this front are known to have led to a dramatic decline in deforestation rates (nearly 70 percent by 2007) (Fearnside 2017). These domestic dynamics, coupled with the Brazilian leadership's aspiration to catapult the country to the position of a 'responsible' global power, led to Brazil agreeing to a legally binding agreement at the 2011 Durban Summit (if the developed countries raised their commitments), to which India and China were opposed.[7]

On the issue of MRV, India proposed the 'International Consultation and Analysis' (ICA) model, based on the ideas of 'non-intrusiveness', 'non-punitiveness', and 'national sovereignty' as a way forward to monitor, report, and verify the Nationally Appropriate Mitigation Actions (NAMAs) (of developing countries) (Goswami 2010). China was reluctant to come on board till the last moment, which could be attributed to its perceptions of sovereignty, transparency, and non-interference that are considered more rigid than India's (Wu 2013, p. 787). South Africa's regional identity ("African Common Position" on climate change) played into its decision to negotiate from the position of first, a regional leader, and then, a global one. After facing backlash from the other African countries for siding with the other emerging economies at the Copenhagen Summit, South Africa had to realign its positions to engage more closely with them, through the African Union, Southern African Development Community, New Partnership for Africa's Development etc. (Masters 2011). This is despite the fact that aligning with the other BASIC countries increased its profile in the climate change negotiations significantly (Hoste 2010).

In the climate change negotiations, the emerging economies have consistently demanded greater level of climate action from the developed countries, based on the historical responsibility of the latter in contributing the largest share of GHG emissions in the atmosphere as well as their historically 'unjust' act of accumulating 'relative' or 'differentiated' capacities to undertake climate action through colonialism, imperialism, and similar other means. The developing countries have used the platform of G-77 to project a united stance on issues such as CBDR-RC, equity, adaptation, as well as financial and technological support. Since poverty alleviation or reduction and development in social and economic sectors are their primary priorities, they have continuously stated that their commitments should be hinged on the principle of equity (equitable allocation of the limited GHG emissions space to meet their developmental needs), and that they would be contingent on support from the industrialised countries. In addition, since the per capita emissions of the majority of developing countries continue to be low, it is morally justifiable to set differentiated goals that do not deprive them of their 'right to develop' (Friman and Hjerpe 2014). As far as climate finance is concerned, the G-77 has traditionally preferred public sources over private sources and has called for a shift from donor-dominated 'assistance formula' to "rights-based resource transfers", also underscoring the need for "new and additional funds" instead of dislodging the existing development assistance (Hochstetler and Viola 2012).

However, at the Copenhagen Summit, the BASIC was seen as distinct from not only the developed countries but also the G-77 to some extent (Bidwai 2014). One could argue that the decision of the BASIC countries to break ranks with G-77 to safeguard their economic interests and the continued framing of climate change as an existential issue by the most vulnerable countries within the G-77 pulled them apart. There have been disagreements on the nature of the climate change agreement itself. At the Copenhagen Summit, the island states and LDCs supported an ambitious, legally binding agreement that would limit the temperature rise to below 1.5 degrees Celsius and would require both the developed countries and emerging economies to act in terms of climate change mitigation (Dimitrov 2010). However, the BASIC countries were in favour of a bottom-up approach that is not necessarily legally binding and ensures less intrusiveness in NAMAs, thereby also insisting on a loose MRV framework. The BASIC countries, primarily India and China, advocated an MRV framework under which all mitigation actions undertaken by non–Annex I parties (COP) would be subject to domestic MRV (with provision for international consultation and analysis). The ones with international support would be subject to international MRV, as stated in the Copenhagen Accord (UNFCCC 2009). This was a part of a bigger package that included greater transparency (mostly insisted by the US) and the Green Climate Fund (agreed upon in 2009 and set up in 2010 within the UNFCCC's framework).[8] The initial aim was to mobilise $100 billion a year by 2020 (that has remained unfulfilled). Moreover, the developing countries have also criticised the fund's dependence on private sources, with the rich countries not committing to enough public sources (Sethi 2011). This itself was a shift in the BASIC countries' climate diplomacy positions – from being largely opposed to any form of international scrutiny to becoming open to the idea of international MRV and/or ICA, especially since they began to realise that the flow of finances and technology was unlikely to come through.[9]

The BASIC countries' positions on the flexibility mechanism of Clean Development Mechanism (CDM) have also seen remarkable shifts. China and India have been the biggest beneficiaries of this mechanism. According to figures, "China hosted more than 49% of total projects between 2004 and 2015, followed by India with 20.6% and Brazil with 4.4%" (Soni and Bhanawat 2018). The CDM was once opposed by developing countries such as India on the grounds that this could put the onus on them to act more on climate mitigation. However, India shifted its position once it realised that it could benefit from CDM and, more specifically, that its industry could get access to external resources and technologies through this mechanism (Sengupta 2012). Even in the case of licensing of clean energy technologies in the developing world, China, India, Brazil, and South Africa are among the top seven beneficiaries with 25 percent, 17 percent, 12 percent, and 3 percent, respectively (as per figures published in 2010 by the UNEP, the European Patent Office, and the International Centre for Trade and Sustainable Development) (EPO 2010). Progress in infrastructure development, extensive human capital, better investment climate, and other favourable factors have, in fact, led to the emergence of countries such as China and India as global leaders in renewable energy, particularly solar and wind energy (UNFCCC 2017). According to figures,

> of the top ten wind power companies in the world, four are Chinese (Sinovel, Goldwind, Dongfang Electric and United Power) and one is Indian (Suzlon). Among the top ten solar energy producers worldwide, seven are from China (LDK Solar, Suntech, JA Solar, Trina Solar, Yingli Green Energy, Hanwha Solar One, and Jinko Solar).
>
> (Abdel-Latif 2012)

Commensurate with their growing GDP, resource base, spending on research and development (R&D), and stature in the international system, the emerging economies have also joined the race for patents and Intellectual Property Rights (IPR), with China as the leading player in terms of patent filing trends in clean energy technology. Interestingly, a country such as India, and even China, had for long advocated for a special provision for green technology under the IPR regime, akin to the pharmaceutical technologies (in the case of HIV/AIDS). On the occasion of the New Delhi High Level Conference on 'Climate Change: Technology Development & Transfer' in 2009, the then Indian Prime Minister Manmohan Singh, stated,

> Such an approach has been adopted successfully in the case of pharmaceutical technologies for the benefit of HIV/AIDS victims in developing countries. The moral case of a similar approach for protecting our planet and its life support system is equally compelling.

> (PMO 2009)

However, India's position has evolved over a period of time; so has China's, in light of its emergence as "the largest hub in renewable energy related components manufacturing, selling, importing, and exporting" (Khurana and Bandyopadhyay 2018).

China, India, and Brazil are known to have increased their R&D investments to advance technological diffusion by enhancing their absorptive capacity. Among the sectors that these countries have concentrated upon for patent filing are solar and carbon capture technologies, hydro/marine and biofuel technology (mainly Brazil), and even geothermal technology (mainly China) (World Energy Council n.d.). Clearly, the mood has shifted somewhat in favour of the IPR regime – from a barrier to an essential incentive for promoting innovation in the field of green technology and developing competitiveness in the global renewable energy market (Lema and Lema 2012). Yet another factor that needs to be taken into consideration is the growing number of bilateral and multilateral engagements and agreements between the emerging economies and industrialised countries (Kasa, Gullberg and Heggelund 2007). The most prominent examples are that of the Asia-Pacific Partnership on Clean Development and Climate, that was flagged off primarily by the US as an alternative to or diversion from the KP, which was also joined by countries such as China and India (Fujiwara 2007). In the 2000s, China signed bilateral agreements on climate change and clean energy with developed parties such as the EU (The Council of European Union 2005) and Australia (Australian

Consulate-General 2007). India also signed several bilateral agreements during this period, including the Partnership to Advance Clean Energy (PACE) with the US in 2009 (US Energy Department n.d.).

The BASIC grouping from Copenhagen to Paris Summits and beyond

The BASIC's fundamental positions underwent changes between the Copenhagen and Paris Summits. These changes were brought about partly by the changing domestic realities and partly by the changing dynamics of the climate change negotiations itself. A change in perspective on how to address climate change – from "how can we share the burden of climate action" to "how can we share the benefits of transition" – is known to have played a big role in influencing the positions of all countries. This shift is more applicable in the emerging economies who had become more keen on sharing the responsibility, but with more benefits than costs. Some would even go to the extent of arguing that during the years between the Copenhagen and Paris Summits, power, or more accurately, 'realist' politics had been overpowered by ideas of 'responsible' leadership, benefits, and co-benefits of transition etc.[10] At the same time, the integration of climate action with the Sustainable Development Goals (SDGs) and the greater involvement of private actors (industries, cities etc.) in deciding the future of the climate change regime have also changed the course of the climate change negotiations.[11]

The differences among the BASIC countries with respect to perceptions of various issues widened during 2009–2019. The solidarity exhibited by them at the Copenhagen Summit failed to sustain the same degree of intensity later on, especially since the Durban Summit in 2011. The BASIC has now been eclipsed by the Like-Minded Group of Developing Countries (LMDCs) that consists of China and India (besides a few other developing/emerging countries) but not Brazil and South Africa. At the Durban Summit, for instance, while India strongly objected to the proposal for adopting a 'protocol' or 'legal instrument' in 2015, both Brazil and South Africa did not oppose it. China also agreed in principle to this proposal due to intensifying pressure from not just developed countries but also developing countries that wanted China to commit more (Minas 2013). The Durban Summit was also characterised by close coordination between the EU and groups such as the Alliance of Small Island States (AOSIS), LDCs, and the Independent Association of Latin America and the Caribbean (AILAC), as corroborated by Paul Watkinson, one of the EU's lead negotiators

Table 2.3 BASIC Countries' Per Capita GDP and Per Capita CO_2 Emissions

Country	GDP Per Capita (2019) (current US$)	CO_2 Emissions Per Capita (2016) (metric tons)
Brazil	8,717.2	2.2
China	10,261.7	7.2
India	2,099.6	1.8
South Africa	6,001.4	8.5

Source: The World Bank.

at the conferences of Copenhagen, Cancun, Durban, Doha, and Warsaw; and chief negotiator and head of the climate negotiations team for the French ministry of Ecology, Sustainable Development and Energy.[12] Essentially, India was isolated. In the end, all the countries, including India, agreed upon including "agreed outcome with legal force" in the outcome of the Durban Summit, "Durban Platform for Enhanced Action" (UNFCCC n.d.).

One also needs to take into consideration the fact that India's idea of equity and justice are linked closely to the per capita principle, which is not the case with the other BASIC countries. If per capita emissions and per capita GDP are taken into consideration, India is far behind the other three countries (Table 2.3).

This is the reason why the per capita principle is so central to India's negotiating positions, which it holds as a legitimate principle on which any climate change regime should be founded. A paper on the conceptual and practical aspects of equity, prepared by experts from the BASIC countries, clearly symbolises this difference between the BASIC countries, with the per capita accumulative principle not being at the centre of approaches adopted by South Africa and Brazil towards equity. If one contextualises equity using these two parameters, its operationalisation in practical terms could lead to outcomes, as explained below (Mattoo and Subramanian 2011):

> Allocations based on GDP per capita favour the poorest countries such as India and Indonesia. It is noteworthy that this principle gives India over two times as much as China in emissions per capita, reflecting China's higher income level. Similarly, India receives about three-and-a-half times emissions per capita as Brazil under the ability-to-pay principle because of the corresponding income differentials, and this translates into even larger differences in total emissions because India's population is about six times greater.

Yet other principles such as historical responsibility, differentiated responsibilities (in light of developing countries' role in sharing the burden of climate change mitigation), and respective capabilities are shared by all. These factors are crucial for the operationalisation of the principle of equity (Lele and Jayaraman 2011). The BASIC countries, in unison, define it as distribution of global carbon budget based on a multilateral agreement about "equitable burden-sharing", including "historical responsibility for climate change", the need to allow developing countries "equitable space for development", and "adequate finance, technology" (South Centre 2010).

A combination of international, regional, and domestic identities and factors distinguish the BASIC countries from each other, some of which have already been touched upon earlier in the chapter briefly. Paul Watkinson states:

> The BASIC never had a truly negotiating position. It was more about political messaging. It is symbolically important for the BASIC countries to have a common ground, which is why there was some level of coordination in terms of formalising ideas around negotiations to find solutions. But in reality, their interests have been different.

Blandine Barreau, a member of the French COP-21 negotiation team in Paris, says that the BASIC countries negotiate for different causes and from varying positions despite sharing common interests in certain issues. As summed up by her, Brazil, with the historical legacy of the Rio Summit and environmental assets such as the Amazon, shows willingness to fulfil its role in terms of what it sees on the international landscape, as long as the agreement would not hurt its autonomy in forest management. Barreau adds that China reached a great deal of the agreement through the joint declaration between China and the US (with reference to the agreement reached by them before the 2015 Paris Summit), and it is now focussed upon decarbonising its economy by leveraging its technological advantage (green technology). South Africa has been playing a difficult game, balancing its role in the BASIC, Africa, and G-77. India is most wary about being coupled with China, as there is a tendency among the European negotiators to think of BASIC as one and as a grouping led by China; and that if one talks to China and gets them on board a decision, others would also follow. India, being aware of this prejudice of the developed countries, while working closely with the other BASIC countries, has not preferred to be seen in the same light as China.[13]

South Africa is always seen to be friendlier towards Europe due to its continued demand for financial and technological support and focus on adaptation (including its efforts to stage itself as the leader of adaptation globally). China and India had slowly begun to wean themselves off from the argument on finance and technology, except using it as a shield against punitive measures (MRV) and legally binding emissions reduction measures. For countries such as India, China, and Brazil, 'historical responsibility' is more important, with China emphasising it most among the BASIC countries. Among other differences, both Brazil and South Africa began to talk about climate action based on the 'peak-plateau-decline (PPD) trajectory', whereas China and India emphasised the co-benefits of climate action more. Some experts believe that India and China have been together all along, despite these differences. Even South Africa cannot take a position that is completely contrary to that of India or China. In any case, the BASIC as a grouping has been successful in resisting international pressure.[14]

The rise of the LMDC in 2012 is indicative of a new dynamic in the international climate order. The LMDC is a much less tight grouping in comparison to the BASIC. In fact, Brazil and South Africa are not a part of LMDC. China and India joined hands with countries such as Bangladesh, Bolivia, Ecuador, Egypt, El Salvador, Indonesia, Iran, Iraq, Malaysia, Mali, Pakistan, Saudi Arabia, Venezuela, and Vietnam among others to advance the 'interests' of developing countries. It was also evident that India began to side more with the LMDC since it could not count even on China to back its demands. It is noteworthy that this grouping consists of struggling economies, rapidly industrialising countries, fossil-fuel-exporting countries, and highly vulnerable countries. Clearly, their interests and world views do not always coincide, but they have been united on the principle of CBDR-RC. The LMDC has worked towards operationalising differentiation across different aspects of the post-2020 climate regime that matter most to the developing countries. At the Paris Summit specifically, it played a major role in working out a compromise formula that underlined the indispensability of 'differentiation' while compromising on 'historical responsibility', which had not found much traction since the 2010 Cancun Summit, as countries had, in principle, reached a consensus to adopt an agreement that is "applicable to all". The LMDC has also been consistently highlighting the non-fulfilment of pre-2020 pledges by the developed countries – something that gained more momentum after the signing of the Paris Agreement, mostly in resistance to the rich countries' demand for show of greater 'ambition' from the developing countries.[15] This group, primarily led by China and India, continues to reinforce the ideas of development and survival.[16]

The rise of the LMDC may not signal deep divisions between the BASIC countries as they both continue to uphold the ideas of equity, historical responsibility, CBDR-RC, climate justice, primacy of the UNFCCC, and so on. Therefore, the LMDC more so signifies the flexibility in the climate order where developing countries choose to leverage their combined strengths through several groups (Hallding et al. 2013). In Chapter 3, the shifts in the BASIC countries' climate diplomacy positions and the drivers of these shifts are analysed individually.

Notes

1 The author's interview with Cleo Paskal, associate fellow in the Energy, Environment and Resources programme and the Asia-Pacific programme, on March 23, 2015, New Delhi.

2 More information on the Ad Hoc Working Group on Long-term Cooperative Action is available on the UNFCCC's website: https://unfccc.int/awg-lca-bodies-page.

3 The equal per capita entitlements approach "first establishes an allowable level of global emissions, termed an emissions budget", which "reflects the ultimate level at which to stabilize GHG concentrations over time, or the amount of GHGs that can be safely emitted in the atmosphere while meeting the ultimate objective of the UNFCCC". It is then "distributed equally among the global population, thereby implying an equal right to the atmosphere, with each country getting an entitlement proportional to its population".". More information can be found in the chapter: Aslam, Malik Amin. 2002. "Equal Per Capita Entitlements: A Key to Global Participation on Climate Change?" In *Building on the Kyoto Protocol: Options for Protecting the Climate*, edited by Kevin A. Baumert. Washington: World Resources Institute.

4 The author's interview with R. R. Rashmi – Distinguished Fellow and Programme Director, Earth Science and Climate Change, The Energy and Resources Institute; India's former principal negotiator for climate change negotiations under the UN Framework Convention on Climate Change; and Special Secretary in the Ministry of Environment, Forest & Climate Change in the Government of India – via Microsoft Teams on August 17, 2020.

5 Based on views gathered from the author's interviews with R. R. Rashmi and Navroz Dubash (Professor, Centre for Policy Research, New Delhi) via Zoom on August 10, 2020.

6 The author's interview with Cleo Paskal.

7 The author's interview with Matías Franchini, Assistant Professor of International Relations at the Universidad del Rosario, Bogotá via Skype on May 31, 2019.

8 The author's interview with R. R. Rashmi.

9 The author's interview with R. R. Rashmi.

10 The author's interview with Navroz Dubash.

11 Inferences from the keynote address by Johan Kuylenstierna, Vice Chair, Swedish Climate Policy Council, at the 4th biannual European Conference of Defence and the Environment (ECDE) in Stockholm, during May 14–16, 2019.
12 The author's interview with Paul Watkinson via Skype on January 17, 2019.
13 The author's interview with Blandine Barreau via Skype on November 20, 2020.
14 Based on views gathered from the author's interviews with Blandine Barreau, R. R. Rashmi, Indrajit Bose (senior researcher, Third World Network, via Skype on July 8, 2020), and Navroz Dubash.
15 The author's interviews with Navroz Dubash, R. R. Rashmi, and Indrajit Bose.
16 The author's interview with Paul Watkinson.

Bibliography

Abdel-Latif, Ahmed. 2012. "Intellectual Property Rights and Green Technologies from Rio to Rio: An Impossible Dialogue?" *International Centre for Trade and Sustainable Development Policy Brief* 14. July 30. Accessed July 20, 2020. https://papers.ssrn.com/sol3/papers.cfm?abstract_id=2120201.

Australian Consulate-General. 2007. "Australia and China - Climate Change and Energy Agenda." *Australian Consulate-General, Guangzhou.* September 11. Accessed July 4, 2020. https://guangzhou.china.embassy.gov.au/gzho/MediaEN28.html.

Bidwai, Praful. 2014. "The Emerging Economies and Climate Change: A Case Study of the BASIC Grouping." *Transnational Institute.* September 4. Accessed July 2, 2020. https://www.tni.org/files/download/shifting_power-climate.pdf.

Chauhan, Chetan. 2010. "India Is World's Third Largest Carbon Emitter." *Hindustan Times.* October 4. Accessed July 2, 2020. https://www.hindustantimes.com/delhi-news/india-is-world-s-third-largest-carbon-emitter/story-wX9gHnxHMcqv4OrownkZ0H.html.

Ciplet, David, Roberts, J. Timmons and Khan, Mizan. 2013. "The Politics of International Climate Adaptation Funding: Justice and Divisions in the Greenhouse." *Global Environmental Politics* 13 (1): 49–66.

Clark, Duncan. 2009. "Growth of Global Carbon Emissions Halved in 2008, Say Dutch Researchers." *The Guardian.* June 25. Accessed July 3, 2020. https://www.theguardian.com/environment/2009/jun/25/carbon-emissions.

Cui, W. J. 2016. "Comparison between North-South Aid and South-South Cooperation: Based on the Analysis of the New Development Finance Institutions." *Journal of Shanghai Jiaotong University* 21 (1): 25–32.

Dasgupta, Saibal. 2009. "Copenhagen Conference: India, China Plan Joint Exit." *Times of India.* November 28. Accessed July 2, 2020. https://timesofindia.indiatimes.com/india/Copenhagen-conference-India-China-plan-joint-exit/articleshow/5279771.cms.

Dimitrov, Radoslav. 2010. "Inside Copenhagen: The State of Climate Governance." *Global Environmental Politics* 10 (2): 18–24.

EPO. 2010. "Patents and Clean Energy: Bridging the Gap between Evidence and Policy." *European Patent Office*. Accessed July 2, 2020. http://documents. epo.org/projects/babylon/eponet.nsf/0/cc5da4b168363477c12577ad00547289/ $FILE/patents_clean_energy_study_summary_en.pdf.

Eurostat. n.d. "Glossary: Group of Twenty (G20)." *European Union*. Accessed July 3, 2020. https://ec.europa.eu/eurostat/statistics-explained/index.php/ Glossary:Group_of_Twenty_(G20).

Fearnside, Phillip. 2017. "Deforestation of the Brazilian Amazon." *Oxford Research Encyclopedias*. September. Accessed July 3, 2020. https://oxfordre. com/environmentalscience/view/10.1093/acrefore/9780199389414.001.0001/ acrefore-9780199389414-e-102.

Friman, Mathias and Hjerpe, Mattias. 2014. "Agreement, Significance, and Understandings of Historical Responsibility in Climate Change Negotiations." *Climate Policy* 15 (3): 302–320.

Fujiwara, Noriko. 2007. "The Asia-Pacific Partnership on Clean Development and Climate: What It Is and What It Is Not." *Centre for European Policy Studies Policy Brief*. November 5. Accessed July 30, 2020. https://www. ceps.eu/ceps-publications/asia-pacific-partnership-clean-development-and-climate-what-it-and-what-it-not/.

G20 Foundation. n.d. "What Is the G20." *G20 Foundation*. Accessed July 3, 2020. https://www.g20foundation.org/g20/what-is-the-g20.

Goswami, Urmi A. 2010. "Ramesh's ICA Proposal Gets Support from BASIC Countries." *Economic Times*. December 8. Accessed July 2, 2020. https:// economictimes.indiatimes.com/news/economy/foreign-trade/rameshs-ica-proposal-gets-support-from-basic-countries/articleshow/7063637. cms?from=mdr.

Gray, Kevin and Gills, Barry K. 2016. "South–South Cooperation and the Rise of the Global South." *Third World Quarterly* 37 (4): 557–574.

Gupta, Joyeeta. 1997. *The Climate Change Convention and Developing Countries: From Conflict to Consensus?* Amsterdam: Springer.

Hallding, Karl, Olsson, Marie, Atteridge, Aaron, Vihma, Antto, Carson, Marcus, and Román, Mikael. 2011. "Together Alone: BASIC Countries and the Climate Change Conundrum." *SEI*. November 28. Accessed August 7, 2020. https://www.sei.org/publications/together-alone-basic-countries-and-the-climate-change-conundrum/.

Hallding, Karl, Jürisoo, Marie, Carson, Marcus and Atteridge, Aaron. 2013. "Rising Powers: The Evolving Role of BASIC Countries." *Climate Policy* 13 (5): 608–631.

Hewitt, Gavin. 2009. "Europe Snubbed in Copenhagen?" *BBC*. December 22. AccessedJuly2,2020.http://www.bbc.co.uk/blogs/thereporters/gavinhewitt/ 2009/12/s_5.html.

Hochstetler, Kathryn and Viola, Eduardo. 2012. "Brazil and the Politics of Climate Change: Beyond the Global Commons." *Environmental Politics* 21 (5): 753–771.

Hoste, Jean-Christophe. 2010. "Where Was United Africa in the Climate Change Negotiations?" *Royal Institute for International Affairs – Africa Policy Brief* 2. February 2. Accessed July 3, 2020. http://www.egmontinstitute. be/where-was-united-africa-in-the-climate-change-negotiations/.

IBSA. 2003. "IBSA - Introduction." *India Brazil South Africa Forum.* June. Accessed July 2, 2020. http://www.ibsa-trilateral.org/about_ibsa.html.

Jaura, Ramesh. 2006. "G8 SUMMIT: Energy Security Yes, but Climate Security Too." *Inter Press Service.* July 10. Accessed July 2, 2020. http://www.ipsnews. net/2006/07/g8-summit-energy-security-yes-but-climate-security-too/.

Jones, Nicola. 2007. "China Tops CO_2 Emissions." *Nature.* June 20. Accessed July 2, 2020. https://www.nature.com/news/2007/070618/full/news070618-9. html.

Kasa, Sjur, Gullberg, Anne T., and Heggelund, Gørild. 2007. "The Group of 77 in the International Climate Negotiations: Recent Developments and Future Directions." *International Environmental Agreements: Politics, Law and Economics* 8 (2): 113–127.

Khurana, Shweta and Bandyopadhyay, T. K. 2018. "Patenting in Renewable Energy Sector - An Analysis." *Journal of Intellectual Property Rights* 23 (1): 44–50.

Kjøllesdal, Kristian and Welle-Strand, Anne. 2010. "Foreign Aid Strategies: China Taking Over?" *Asian Social Science* 6 (10): 3–13.

Lele, S. and Jayaraman, T. 2011. *Ministry of Environment & Forests, Government of India.* April 13. Accessed August 17, 2020. http://atree.org/sites/ default/files/pubs/slele/working-papers/2011-05-1-GSP-Note-Equity-in-Context-of-Sustainable-Development-INDIAv2.pdf.

Lema, Rasmus and Lema, Adrian. 2012. "Technology Transfer? The Rise of China and India in Green Technology Sectors." *Innovation and Development* 2 (1): 23–44.

Masters, Lesley. 2011. "Sustaining the African Common Position on Climate Change: International Organisations, Africa and COP17." *South African Journal of International Affairs* 18 (2): 257–269.

Mattoo, Aaditya and Subramanian, Arvind. 2011. "Equity in Climate Change: An Analytical Review." *World Bank Group Policy Research Working Paper.* Accessed July 4, 2020. https://elibrary.worldbank.org/doi/ pdf/10.1596/1813-9450-5383.

Minas, Stephen. 2013. "BASIC Positions-Major Emerging Economies in the UN Climate Change Negotiations." *Foreign Policy Centre Briefing.* July 5. Accessed July 31, 2020. https://fpc.org.uk/fpc-briefing-basic-positions-major-emerging-economies-in-the-un-climate-change-negotiations/.

Mishra, Abhishek. 2018. "IBSA and South-South Cooperation: An Appraisal." *Observer Research Foundation.* June 20. Accessed August 20, 2020. https:// www.orfonline.org/expert-speak/ibsa-and-south-south-cooperation-an-appraisal/.

NEAA. 2008. "Global C02 Emissions from Fossil Fuels and Cement Production by Region, 1990–2007." *Netherlands Environmental Assessment Agency.* September 30. Accessed July 31, 2020. http://www.mnp.nl/en/ publications/2008/GlobalC02emissionsthrough2007.html.

Obama, Barack. 2009. "Major Economies Forum on Energy and Climate." *US Department of State.* July 9. Accessed July 2, 2020. https://2009-2017. state.gov/e/oes/climate/mem/index.htm.

Paskal, Cleo. 2010. "How Copenhagen Has Changed Geopolitics: The Real Take-Home Message Is Not What You Think." *New Security Beat.* January 4. Accessed July 2, 2020. https://www.newsecuritybeat.org/2010/01/how-copenhagen-has-changed-geopolitics-the-real-take-home-message-is-not-what-you-think/.

PMO. 2009. "PM's Address at the New Delhi High Level Conference on 'Climate Change: Technology Development & Transfer'." *Prime Minister's Office, India.* October 22. Accessed July 3, 2020. https://archivepmo.nic.in/ drmanmohansingh/speech-details.php?nodeid=803.

https://www.thehindu.com/news/international/Bangladesh-disappointed-by-draft-treaty-framed-by-India-China/article16852169.ece.

Rapp, Tobias, Schwägerl, Christian and Traufetter, Gerald. 2010. "How China and India Sabotaged the UN Climate Summit." *Spiegel.* May 5. Accessed July 3, 2020. https://www.spiegel.de/international/world/the-copenhagen-protocol-how-china-and-india-sabotaged-the-un-climate-summit-a-692861.html.

Republic of Indonesia. 1955. "Final Communiqué of the Asian-African Conference of Bandung (24 April 1955)." *The Ministry of Foreign Affairs.* Accessed July 31, 2020. http://www.cvce.eu/obj/final_communique_of_ the_asian_african_conference_of_bandung_24_april_1955-en-676237bd-72f7-471f-949a-88b6ae513585.html.

Rong, Fang. 2010. "Understanding Developing Country Stances on Post-2012 Climate Change Negotiations: Comparative Analysis of Brazil, China, India, Mexico, and South Africa." *Energy Policy* 38 (8): 4582–4591.

Scott, Shirley V. 2018. *Climate Change and the UN Security Council.* Cheltenham: Edward Elgar.

Sengupta, Sandeep. 2012. "International Climate Negotiations and India's Role." In *Handbook of Climate Change and India: Development, Politics and Governance,* edited by Navroz K. Dubash, 101–117. New York: Routledge.

Sethi, Nitin. 2011. "A Green Climate Fund but No Money at Durban." *The Times of India.* December 6. Accessed August 3, 2020. https://timesofindia. indiatimes.com/home/environment/developmental-issues/A-green-climate-fund-but-no-money-at-Durban/articleshow/11009614.cms.

Soni, Monika and Bhanawat, Shurveer S. 2018. "Accounting and Taxation Issues of Carbon Credit Transactions." *Pacific Business Review International* 10 (12): 41–50.

South Centre. 2010. "Ministers of Basic Countries (Brazil, China, India, South Africa) Stress Primacy of United Nations as Forum for Climate Negotiations and that the World Cannot Wait Indefinitely for the United States to Act on Climate Change." *South Centre.* April 27. Accessed July 3, 2020. https://www.southcentre.int/wp-content/uploads/2013/08/ PR03_BASIC-Ministerial-statement-on-climate-change_EN.pdf.

The Council of European Union. 2005. "Joint Declaration on Climate Change between China and the European Union." *European Union.* September 7. Accessed July 3, 2020. http://register.consilium.europa.eu/doc/srv?l=EN&f=ST%2012009%202005%20INIT.

The Guardian. 2009. "Draft Copenhagen Climate Change Agreement - the 'Danish Text'." *The Guardian.* December 8. Accessed July 3, 2020. https://www.theguardian.com/environment/2009/dec/08/copenhagen-climate-change.

The Hindu. 2009. "Jairam Ramesh Statement on Copenhagen Accord in Rajya Sabha." *The Hindu.* December 22. Accessed July 2, 2020. https://www.thehindu.com/news/national/Jairam-Ramesh-statement-on-Copenhagen-Accord-in-Rajya-Sabha/article16855083.ece.

The Third Pole. 2015. "A Divided South Asia at COP21." *The Third Pole.* December 1. Accessed July 2, 2020. https://www.thethirdpole.net/2015/12/01/a-divided-south-asia-at-cop21/.

The White House. 2014. "U.S.-China Joint Announcement on Climate Change." *Office of the Press Secretary.* November 12. Accessed July 24, 2020. https://obamawhitehouse.archives.gov/the-press-office/2014/11/11/us-china-joint-announcement-climate-change.

UN. 1974. "Declaration on the Establishment of a New International Economic Order." *United Nations Documents.* May 1. Accessed July 3, 2020. http://www.un-documents.net/s6r3201.htm.

UNFCCC. n.d. "Essential Background - Durban Outcomes." *United Nations Framework Convention on Climate Change.* Accessed July 3, 2020. https://unfccc.int/process/conferences/the-big-picture/milestones/outcomes-of-the-durban-conference.

UNFCCC. 2008. "Report of the Conference of the Parties on Its Thirteenth Session, Held in Bali from 3 to 15 December 2007." *United Nations Framework Convention on Climate Change.* March 14. Accessed July 3, 2020. https://unfccc.int/resource/docs/2007/cop13/eng/06a01.pdf.

UNFCCC. 2009. "Copenhagen Accord." *United Nations Framework Convention on Climate Change.* December 18. Accessed July 2, 2020. https://unfccc.int/resource/docs/2009/cop15/eng/l07.pdf.

UNFCCC. 2017. "China and India Lead Global Renewable Energy Transition." *United Nations Framework Convention on Climate Change.* April 21. Accessed July 2, 2020. https://unfccc.int/news/china-and-india-lead-global-renewable-energy-transition.

UNSC. 2007. "Security Council Holds First-Ever Debate on Impact of Climate Change on Peace, Security, Hearing over 50 Speakers." *United Nations.* April 17. Accessed July 3, 2020. https://www.un.org/press/en/2007/sc9000.doc.htm.

US Energy Department. n.d. "U.S.-India Energy Cooperation." *US Energy Department.* Accessed July 3, 2020. https://www.energy.gov/ia/initiatives/us-india-energy-cooperation.

World Energy Council. n.d. "Energy Sector Environmental Innovation: Understanding the Roles of Technology Diffusion, Intellectual Property

Rights, and Sound Environmental Policy for Climate Change." *World Trade Organization.* Accessed July 2, 2020. https://www.wto.org/english/forums_e/ngo_e/wec_rules_2011_e.pdf.

Wu, Fuzuo. 2013. "China's Pragmatic Tactics in International Climate Change Negotiations: Reserving Principles with Compromise." *Asian Survey* 53 (4): 778–800.

Zheng, Chen. 2016. "China Debates the Non-Interference Principle." *The Chinese Journal of International Politics* 9 (3): 349–374.

3 Country-wise analysis of BASIC's climate diplomacy positions

China: a global climate leader?

Apart from external pressure, and growing geopolitical and geoeconomic clout, the domestic situation and leadership have had a huge role to play as well in shaping China's climate diplomacy after the Copenhagen Summit. Politically, China continued to side with India in the BASIC grouping, but fissures began to emerge as China also began to coordinate its positions with the US and the EU, culminating in a joint communiqué with the US. After months of negotiations between the US and China – the world's largest GHG emitters – this agreement was signed in 2014, which was regarded by many analysts as a "landmark" and/or "historic" one. Both countries decided to move away from business-as-usual. The US announced that it would reduce GHG emissions by 26–28 percent from 2005 levels by 2025. China announced its goal of peaking emissions by 2030 (albeit without specifying the level) as well as its goal to increase the share of non-fossil fuels in primary energy consumption to around 20 percent by 2030 (Landler 2014). Previously, the US had pledged to reduce GHG emissions by 17 percent by 2020 and 83 percent by 2050 from 2005 levels. China had agreed to reduce carbon emissions per unit of GDP by 40–45 percent from 2005 levels by 2020, besides boosting the share of non-fossil fuels to around 15 percent by the same year (Schmidt 2010). This agreement was considered a major shift from China's previous positions, as it blurred the line of differentiation between China and the industrialised countries, something that China did not compromise on since the beginning of the negotiations.

When President Xi Jinping took the mantle of leadership in China in 2012, he made his intention of raising China's position to that of a pivotal player in global governance, including on climate change, clear. China positioned itself to seek "compromises with both the EU and, particularly, the US in order to demonstrate leadership" (Belis et al. 2018). According to Xiaosheng (2018), China transitioned from a "dead weight" or a "wreck", as per external perceptions of China's role in the negotiations, to a "global climate leader" at the 2015 Paris Summit. While China's change in climate diplomacy positions could be attributed to international pressure and its growing confidence, the domestic challenges related to pollution are also known to have played a crucial role in its decision to move away from traditionally held positions. The worsening pollution levels have been singled out as the cause for deteriorating health, economic losses, poorer industrial efficiency, and even lower agricultural production in several parts of the country (The Economist 2013). Increasing discontent among the citizens and recurring economic havoc due to air pollution have forced the Chinese government to act more "responsibly" at the national and international levels (Luft 2013). Clearly, the Chinese government has drawn a parallel between rapidly increasing coal consumption in the country, worsening pollution levels, and opportunities for climate change mitigation by addressing the problem of pollution. Realising the "co-benefits" of climate action, China decided to tackle the issue of pollution by focussing on steps that could potentially reduce its excessive dependence on coal, thereby also aiding climate change mitigation, contributing to global climate action efforts (Li 2016).

Although China has rarely identified itself as a "vulnerable" country at the international level, it has been in the process of studying climate vulnerabilities and their social and economic impacts (Hung and Tsai 2012). Apart from pollution control, "structural economic reform to boost cleaner industries", "development of strategic renewable industries", "international reputation", "access to markets", and "acquisition and development of advanced low-carbon technologies" have been identified as the major co-benefits (Gallagher and Xuan 2019, p. 3). Furthermore, President Xi Jinping reintroduced the idea of "ecological civilisation" (first appeared in 2007), aimed at implementing environmental laws and reforms to large-scale environmental problems in the country that arose from unhindered developmental activities until then. Interestingly, Xi invokes the concept of national security in order to raise the urgency of environmental problems that the country faces in an article published in 2019. He calls for "effective prevention of ecological and environmental risks, as ecological and environmental security is an important part of national security, as well as an important guarantee to achieve sustainable and healthy development of the Chinese economy and society" (Sheng 2019).

With growing emissions, China had to become more flexible. Hence, international pressure cannot be delinked from China's vision of being acknowledged as a responsible global power. At the same time, at the domestic level, its climate policy process is regarded as "strategic pragmatism", wherein the country's political system dominated by the Communist Party of China (CPC) becomes the key to institutionalisation and augmentation of climate action. Domestically, China has followed a top-down approach, wherein the subnational governments and authorities are entrusted with various responsibilities of implementing policies put in place by top-level authorities. The long-term strategic nature of Chinese decision-making is known to be contingent on its political system that does not depend on elections and terms of leaders. In this context, one would also argue that the Chinese positions are driven by pragmatism and not by any kind of ideology. In fact, the communist ideology is merely "symbolic" and not "motivational" (Gallagher and Xuan 2019, p. 159).

In this context, it may also be important to highlight China's objective of circumventing the "Malacca dilemma", which has always concerned the leadership. Its indispensable dependence on the Malacca Strait for fossil fuel supply and, in turn, energy security has been well documented. Any threat to this strait (piracy, blockade etc.) could put China's surging economic growth in jeopardy. Therefore, China has been investing not only in domestic coal production and gas pipelines but also in renewable energy (Singh 2013). Furthermore, China had begun to view climate change mitigation as a business "opportunity" rather than a hindrance to development, so much so that the rhetoric of "development rights" had given way to that of "responsible major power" in official statements and documents (Jing 2015). The then US President Barack Obama acknowledged the "important role China played in securing a historic climate agreement in Paris" (Reuters 2015). China has also been interested in translating its climate actions into soft power in developing countries, particularly in Asia and Africa. It has already established a climate fund worth 20 billion yuan (approximately US$2.9 billion) to assist developing countries in climate change mitigation and adaptation (Liu 2017).

In the early 2010s, China emerged as the global leader in renewable energy, surpassing the US, Germany, and other developed countries. With the aim of tackling pollution and achieving energy security, China has accelerated investments in the renewable energy sector. According to figures released by the International Energy Agency (IEA), "China accounted for nearly half of all new global capacity in 2017" and "China will account for over 40% of global renewable capacity expansion from 2018 to 2023" (Sugiuira and Okutsu 2018). These developments are being seen by the Chinese establishment as an opportunity to expand its low-carbon industry/technology in other countries – an engine for its economic growth. As stated in a report of the Institute for Energy Economics and Financial Analysis, "China increased its foreign investment in renewables by 60 percent to reach a record $32 billion" in 2016 (Jaeger, Joffe and Song 2017).

China's position in the international climate order has received more credibility after US President Donald Trump's announcement to withdraw from the 2015 Paris Agreement. The shift in China's positions after Trump's ascendancy can be seen to have two dimensions. First, the peer competition and geopolitical rivalry between China and the US spiralled upward, evidenced mainly by the "trade war" on the economic front (SCMP Reporters 2020). Second, President Trump's repeated claims that the reason for his decision is that the Paris Agreement signifies an "unfair" deal from the US's perspective in light of what countries such as China and India could gain out of it (PTI 2018) nullified the cooperative outcomes reached during the Obama era. Bilaterally, many institutional and informal mechanisms set up during the Obama administration have been discontinued. Apart from the closure of the US Office of International Climate and Technology, which has played a key role in forums such as the Clean Energy Ministerial (that includes the US and China among other countries), the China–US Climate Change Working Group (CCWG) operations have also been suspended (Jiahan 2018).

Interestingly, the discourse on Chinese leadership on climate change began to gain more momentum after Trump's announcement. The fear of the global momentum on climate action slowing down was imminent. In such a scenario, many analysts felt that China was now "poised for leadership" and to fill the "leadership vacuum" left by the US. The onus on China to lead in the negotiations, representing various groups, became more apparent. China's climate cooperation with the EU, including in reinforcing the "rule of law" and upholding the Paris Agreement, grew manifold (Kaneti 2020). In fact, as per an EU delegate, "If we get China, the rest of Asia will follow". Furthermore, China's decisions to halt the construction of several coal-powered plants and provide stimulus to renewable energy projects have defined this discourse of a reinvigorated role that it could play in the international climate order (Nunez 2017). More importantly, President Xi's attempts to project his country as the vanguard of globalisation and anti-protectionist values (as opposed to President Trump) began to redefine China's position in the international climate change negotiations as well. China gradually began to distance itself from the dogmas associated with a developing country status and indirectly highlighted the way in which it has benefitted from globalisation (Smith 2017).

The discourse is, however, not straightforward. As some experts comment, China's desire to take on the mantle of climate leadership ebbed in 2018–2019 with the slowing economy and US–China trade war. The discourse of Chinese leadership seems to have given way to the relevance of "multilateralism". China's plans to build coal-powered projects that account for 40 percent of the world's total as well as to finance such projects in Southeast Asia, West Asia, and Africa through the Belt and Road Initiative (BRI) have created a dent in China's image as a "torchbearer" in climate action (Lehr 2019). In essence, major shifts in China's perceptions of its vulnerabilities, position in the international system, and economic and global aspirations changed its climate diplomacy positions substantively – from a naysayer to a proactive player, and balancing between its status of a developing country and that of the world's second-largest economy (after the US). From tackling pollution to diversifying energy supply, China's climate diplomacy has evolved over a period of time, but major shifts continue to be stymied by material (geopolitical and geoeconomic) interests.

Kaneti (2020) analyses China's use of the stratagem – "exchange the roles of host and guest" – to position itself in the international climate order as a leader through several steps. Besides the Trump factor, what one needs to recognise is the way in which China has used several opportunities in this respect. There has been an upsurge in the "moral" influence of China, owing to its massive "green financing" initiatives not only at the domestic level but also in other countries. By becoming the world's "second largest green bond issuer" and investing in the "world's largest national carbon trading scheme", China's place in the "global carbon market", and "green finance and investment" is emboldened. China leaves no stone unturned as far as championing the cause of the developing countries is concerned, whether it is through invoking the CBDR-RC principle or climate justice (Kaneti 2020). In fact, China openly called Australia a "condescending master" in the latter's attitude towards the climate-vulnerable Pacific island countries (Doherty 2019). Here again, China's attempts to showcase its moral uprightness by bringing forth its policy of providing financial assistance without any strings (political) attached takes centre stage. The fact that China wishes to make geostrategic inroads into regions such as the South Pacific cannot be ignored either (Paskal 2020).

In short, China's climate diplomacy is often equated with its geopolitical and geostrategic positioning – in congruence with its "great power" status (Li and Ivleva 2019) and its stated ambition to be a superpower (most often referred to as a "revisionist" one). Climate diplomacy is yet another platform that portrays China's dichotomous approach towards the international system – that can be understood through the interaction between ideational and material forces. On the one hand, China projects itself as a revisionist power, wanting to change the existing rules and norms of the international climate order. On the other hand, it still holds on to many principles of the current status quo that have helped it attain the present status. What one could infer from Chinese positions is that even while it asks for change in the status quo, it never presses for a sweeping or deep-seated one.

Brazil: ebbs and flows in climate diplomacy

Having hosted the first Conference of Parties of the UNFCCC, Brazil has played a major role in the climate change negotiations since the beginning, from leading the G-77 to offering proposals on operationalisation of principles such as "historical responsibility" and facilitating the formation of the CDM through its inputs. Like the other emerging economies, Brazil has perceived climate change as a development issue. While in the climate change negotiations Brazil has aspired to be a climate leader, at the domestic level, Brazil's actions are regarded as less than satisfactory. For instance, Viola and Franchini (2017) observe that during 1990–2004, Brazil was a "climate villain"; during 2005–2010, it adopted a moderately conservative approach in terms of climate commitment; and during 2011–2017, it relapsed to become "climate-negligent". Brazil's position in the international climate order has, therefore, seen more downs than ups, which is now being manifested more evidently in the form of statements being made by its President, Jair Bolsonaro (2019), portraying his ambivalence on the Paris Agreement. Matías Franchini, co-author of *Brazil and Climate Change: Beyond the Amazon*, comments[1]:

> First, Brazil has tried to portray itself through the frame of its climate commitments, especially under President Luiz Inácio Lula da Silva, by doing more than its fair share of responsibility. Second, it has also projected itself as a climate power and leader, and in this respect, owing to the impact of the Amazon deforestation on the carbon cycle and consequent commitments, it is a major actor in the UNFCCC negotiations, thereby declaring its intent of doing, and not just talking. Third, its projection of itself as a champion of the developing world and rigid defender of the

CBDR-RC principle has been a major feature of its climate diplomacy. Fourth, it seeks to act as a bridge between the developed and developing countries. However, the staging of Brazil as a climate leader and power is merely a myth as during 2011–2015, climate action had virtually disappeared at the national level. Moreover, for Brazil, the Latin American identity is important only in the domain of economic integration, and not climate change.

Brazil's perennial problem is deforestation. Brazil "possesses the largest forest carbon and biodiversity stocks of any country in the world" (Viola and Franchini 2012). From opposing the inclusion of forests in the CDM for the fear that the burden would fall upon countries such as Brazil to reduce deforestation, instead of the developed countries that have a historical responsibility for climate change, Brazil moved towards proposing "a global fund for forest conservation", when deforestation rates began to fall in the country. Holding the idea of sovereignty tightly, Brazil has been wary of losing sovereign control over the resources of the Amazon on account of its internationalisation. In fact, this has been an integral part of the Brazilian identity (Kiessling 2018).

Therefore, both security and economic reasons have dictated Brazil's climate diplomacy positions at the international level, as far as its vast forest resources are concerned. However, Brazil, like China, saw an opportunity in leveraging the market mechanisms such as the CDM and REDD-plus (Reducing Emissions from Deforestation and Forest Degradation). In 2008, Norway decided to pay US$1 billion to Brazil's Amazonian Fund, which was set up by the then President Lula (in 2008) by executive order and administered by the Brazilian National Development Bank (BNDES) (Reuters 2015). Similarly, Brazil also saw an opportunity in the form of a global ethanol market, driving the leadership to promote ethanol diplomacy, especially in other emerging economies such as China and India. Brazil's ethanol diplomacy is equally guided by its objective of "fostering a solid international market structure for ethanol", which would strengthen its position as the leader of "global governance of ethanol" and champion of "sugarcane ethanol producers" in developing countries. Furthermore, it would promote the ideas of "energy dependence and rural development" (Farias 2015).

Interestingly, the emergence of domestic groups and coalitions as well as national leaders that favoured strong climate action also steered Brazil to the proactive side of the international climate debate. Reduced deforestation rates, incentivisation of forest protection, and other such measures were already being supported and advanced by the Brazilian NGOs. Some of these changes were brought about by individual leaders who had a special interest in ensuring that the Amazon rainforests were protected. One of them was Marina Silva, who was the Minister of the Environment (2003–2008) in the Lula administration. A staunch defender of the Amazon, Silva is a native Amazonian, and she implemented several measures to ensure close co-ordination between the federal government and the Amazonian state authorities through real-time monitoring of the results, at times even imprisoning employees from these authorities for violation of the law. However, she eventually resigned from the government due to pressure from the biofuels/agribusiness industry and the hydroelectric power industry – both responsible for high deforestation in the Amazon (Grudgings 2008).

Silva also recorded 19 percent of the valid vote in the 2010 Presidential elections, which in itself was considered historical for a pro-environment candidate (Boadle 2014). President Dilma Rousseff too continued to support these industries, albeit due to increasing domestic pressure (also signalled by the support received by Silva), she decided to pay more attention to climate change issues and pledge targets (as she did at the Copenhagen Summit). In addition, the rise of coalitions such as "Open letter to Brazil about climate change" (consisting of 22 large national corporations), "Alliance of Corporations in Favor of the Climate" (mainly agribusiness corporations), and "The Coalition of Corporations for Climate" (utilities and energy corporations) altered the domestic scenario substantially, as these coalitions pushed for a new responsible climate policy (Hochstetler 2012).

Moreover, in 2009, Brazil, under the leadership of President Lula and Minister of the Environment Carlos Minc (another environmentalist and one of the founding members of the Green Party in Brazil), announced that it might consider donating the equivalent of about US$ 5 billion to poor countries, especially in Latin America, for climate change mitigation and adaptation (Ministério do Meio Ambiente 2009). Stemming from the need to not only cater to the demands of the domestic audience but also establish leadership at the international level, Brazil has taken several steps to remodel its climate diplomacy strategy. The fact that the Ministry of Environment began to lead the negotiations from the Brazilian side, instead of the Foreign Ministry and the Ministry of Science and Technology, also resulted in a change in Brazilian stance on climate change. Brazil, which had for long held on to the ideas of sovereignty – based on the right to use its resources to promote development – began to show greater inclination towards international cooperation after the Copenhagen Summit. It was the only large developing country to adopt an absolute economy-wide target or, in other words, an absolute reduction in emissions in the Paris Agreement (Yeo 2015).

In 2012, Brazil was one of the few countries (especially among the developing countries) to witness a significant drop (41 percent) in GHG emissions since 2005 (Timperley 2018), due to its deforestation policies, although the emissions have risen since then due to a surge in illegal deforestation and economic crisis. At the same time, if one discounts deforestation, Brazil's carbon intensity grew during the same period, owing to an increase in diesel consumption, electricity produced from fossil fuels, and oil refining (Viola and Franchini 2012). As far as Brazil's positions on the CBDR-RC and historical responsibility are concerned, they shifted gradually – while President Lula wished to break away from them, in support of Brazil as a climate leader, President Rousseff went back to the "conservative positions" in 2012. During President Rousseff's tenure, "conservative forces" endorsed by the government started to take matters into their own hands, sidelining the "reformists" in the process (Viola and Franchini 2014). For President Rousseff, the Paris Summit became rather important to reinforce her position domestically, as she was in trouble by then due to corruption allegations and economic recession. These developments partially explain the shift in Brazil's climate diplomacy positions at the Paris Summit, as observed by Franchini.[2]

Nevertheless, Brazil gradually began to distance itself from China and India for the fear of being isolated in the climate change negotiations. Incidentally, at the Durban Summit, Brazil was the first country to join hands with the EU and the other countries – as a part of a "high ambition coalition" – leaving behind its BASIC partners, and willing to strike a legally binding climate pact. China and India were not keen on agreeing to the principle of "global stocktake" and India specifically wanted this to be voluntary (Pashley 2015). At the summit, the Brazilian negotiator went on to say that the BASIC was not a negotiating group (Tabau and Lemoine 2012). Brazil is different from India and China, in the sense that it can "do a lot" by focussing on land use and agriculture (Pashley 2015). These disagreements led to the splintering of the BASIC group, with these countries drifting back to the G-77 (on mitigation) as their primary support and negotiating base. In the run-up to the Paris Summit, Brazil made further announcements to fortify its position as a potential climate leader. It committed to increase the share of renewable energy in its electricity generation mix in a joint statement with the US, which also paved the way for a Brazil–United States Joint Initiative on Climate Change, aimed at increasing cooperation on sustainable land use, clean energy, and adaptation (The White House 2015).

In essence, Brazil set the tone for a successful Paris Summit by engaging bilaterally with a few developed countries, declaring an economy-wide target and concurrently allying with the BASIC and G-77 to push the agenda on differentiation based on equity and right to development. It thus, positioned itself to be building bridges between different groups of countries. It is a different matter that the political turmoil in Brazil since the Paris Summit due to several reasons, including the impeachment of President Rousseff, has put Brazil's rising status in the international order in jeopardy; so has it overturned its image as a proactive player in the climate change negotiations. Another important point to be raised here is Brazil's "regional" identity that has been non-existent, according to most experts, which is perhaps in congruence with the misperceptions built around Brazil's climate leadership too.[3] Brazil is not a part of either Bolivarian Alternative for the Americas (ALBA) or AILAC (Watts and Depledge 2018). Despite shared vulnerabilities and Brazil's self-proclaimed leadership on climate change from time to time, it has chosen to negotiate as a part of the BASIC or as a "lone wolf". Brazil has attempted to take an independent position, partly with an ambition to question the existing status quo in the international system, and partly to sustain its economic agenda that it does not wish to compromise for environmental and climate gains (Edwards and Roberts 2015).

Brazil has entered a new era of climate diplomacy under President Bolsonaro, who is often equated with President Trump as far as policies on many issues, including climate change, are concerned. He has not only openly promoted agribusiness activities and hydroelectric projects in the Amazon rainforests but also crippled the funding of NGOs that have actively propagated environmental protection and conservation in the region (Reuters 2019). Since he came to power, deforestation rates have soared. The 2019 Amazon fires were blamed upon "illegal loggers, clandestine miners and aggressive farming businesses" who are reportedly being supported by the Bolsonaro administration (AFP 2019). Interestingly, when French President Emmanuel Macron called for talks on the "international crisis" in the Amazon and made the EU–Mercosur (South American nations) trade deal contingent on Brazil's commitment to climate action, President Bolsonaro reduced his offensive against the Paris Agreement (News Wires 2019) – which he had opposed during the election campaign.

In addition, President Bolsonaro has shut down departments that deal with deforestation and climate change within the Ministry of Environment. His administration also decided not to host the 2019 COP (Escobar 2019), which was later held in Madrid. The period between 2015 and 2019 has been characterised by a wave of anti-science bias of the establishments in many countries, including Brazil (Nature 2018). The ideational dimensions of sovereignty and anti-colonialism continue to exist in the Brazilian discourse on climate action, as evidenced by President Bolsonaro's accusation against French President Macron's alleged "colonial mindset inappropriate in the 21st century" (Schoenwalder 2019). The simmering tension between Brazil's domestic interests and international image that was visible during the Rousseff administration gave way to a more open defiance of the international climate order.

The geopolitical dynamics have brought Brazil and the US closer, while pitting the former against the EU. In addition, Brazil is now keener to cooperate more closely with the other BASIC countries in respect to retaining its old carbon credits or emission reduction certificates (CERs) under the Kyoto regime. At the 2019 Madrid Summit, the BASIC countries sought an "appropriate transition of the CDM" to the new market mechanism under the post-2020 climate regime for "securing continued engagement of the public and private sectors in mitigation action" (PIB 2019). The EU, LDCs, CVF, and several other parties do not want the credits to be carried forward due to the flawed system of CDM, problems of double counting, and risks associated with the flooding of the global carbon market by cheap credits, potentially harming the prospects of additional emissions reduction by countries (Evans and Gabbatiss 2019; Timperley 2019). With these developments, Brazil's previous attempts to demonstrate leadership have largely been eroded.

South Africa: upholding the Africa position

South Africa has been the most forthcoming country when it comes to readiness and willingness to accept a legally binding agreement (Sethi 2010), with the condition that the developed countries would act and that they would provide technological and financial resources to the developing countries. Keeping its "overriding national priorities" of "sustainable development, job creation and poverty eradication" on the forefront, South Africa has always given priority to its national interests while acting on climate change (Kotzé et al. 2016). Even while signing the Paris Agreement, it highlighted the importance of equity and the need for "a multilaterally agreed equity reference framework" to determine its carbon budget (Steenkamp and Naude 2018). What differentiates South Africa from the other BASIC countries (even Brazil) is its explicitly defined regional identity, which puts it at the centre of the African common position on climate change. South Africa, as a part of this common position, has been one of the most vociferous champions of developing countries' right to access "public" resources allocated by developed countries (African Union Development Agency n.d.). Its position in the Africa Group came under the scanner after the Copenhagen Summit for allying with the other BASIC countries and opting for a less-than-satisfactory Copenhagen Accord. Since then, South Africa has been straddling the fence, with greater leaning towards the Africa Group than the BASIC.

South Africa's climate diplomacy has been influenced by its foreign policy agenda – realigning its relations in the post-Apartheid scenario with the West, and at the same time, reinforcing its partnerships with the African and major developing countries based on identity and developmental concerns. As analysed by Rennkamp and Marquard (2018), South Africa has joined various "technical", "carbon pricing", and "great power" climate clubs to achieve its interests, particularly in the arena of climate change mitigation. It aligns itself globally by balancing between the emerging economies and African countries – which is in consonance with its national interests. By adopting an "absolute peak, plateau and decline greenhouse gas emissions trajectory range" with the starting point as 2020, it also balances between its concerns for equity at the international level and a pragmatic commitment (UNFCCC 2015). On the technical aspects, South Africa has mainly attempted to advance knowledge on specific issues concerning review, transparency, mitigation etc. Therefore, South Africa seems to see greater potential for cooperation through technical climate clubs than political coalitions, as seen in the cases of "Partnership for Market Readiness" and "Partnership for Transparency" under the Paris Agreement. Here again, there is greater focus on partnerships between developing countries with "shared experiences of policymaking and implementation" (Rennkamp and Marquard 2018, p. 457).

It cannot be denied that South Africa's profile in the UNFCCC was boosted notably after it joined hands with the BASIC countries as an African powerhouse that could decide the future of the international climate regime, which was however, short-lived. South Africa has realised that in order to push ahead its agenda for development and economic growth and put pressure on the developed countries, it needed to side with the BASIC partners on occasion, as it would not get much leeway on this front regionally (Hallding et al. 2011). Having been recognised as a major (emerging) economy through other forums such as Major Economies Forum (MEF), BRICS, G-20 and so on, South Africa had already begun to be seen in a different league, in contrast to the other African countries, helping it project itself as a stable economy to some extent. Furthermore, South Africa's leadership in the international climate order has also been questioned. Moreover, as some observe, "South Africa's midwife role between major polluters has the potential to resolve the crisis by bringing together the two opposing factions (that is, the USA and China)" (Amusan and Olutola 2016).

South Africa's biggest challenge is its heavy dependence on coal production and consumption, with over 90 percent of electricity generated from coal (UNSTATS 2008). South Africa is the world's fourth-largest coal producer, and according to the available figures, 28 percent of its coal is exported (as of 2013) (Stats SA n.d.). Therefore, as long as there is global demand for coal, it will remain a mainstay of the South African economy and it will stick to "self-seeking climate agenda (if any)" (Amusan and Olutola 2016). Furthermore, powerful lobbies of coal-related industries have been instrumental in delaying longstanding demand for changes at the institutional level that could usher low-carbon development, including introduction of carbon pricing mechanisms (Rennkamp 2019). Like the other BASIC countries, South Africa also announced a series of policies before the Copenhagen Summit, such as the White Paper on Renewable Energy (2003), National Climate Change Response Strategy (2004), and Long-Term Mitigation Scenarios (2007). In addition, the country's National Planning Commission has already initiated a Stakeholder Dialogue on Pathways for a Just Transition in order to prepare the economy and people for a transition to low-carbon economy, considering that the coal mining industry is the third-biggest employer in the country (Stats SA n.d.).

What add to South Africa's challenges are its "low growth potential, slow private investment growth and weak integration into global value chains" (World Bank 2018), contrary to the stories of China, India, and even Brazil to a large extent. Clearly, the stakes for the country are high, as far as climate change mitigation and adaptation are concerned. Yet again, unlike the other African countries, for South Africa, mitigation is as consequential as adaptation, which in the case of the former is largely tilted towards adaptation. This is the reason why South Africa has steadfastly been demanding financial and technological support from the developed countries – reaffirmed in a majority of speeches by the former President Jacob Zuma. A transition to a low-carbon economy, based on the principles of equity and justice at the domestic level, is of utmost importance to South Africa.

The 2011 Durban Summit, once again, proved to be a gamechanger in South Africa's climate diplomacy as well. Since the country hosted the summit, it was upon its leadership to showcase the image of a "responsible" player. South Africa was looking for a comprehensive and ambitious outcome at this summit. Hence, it formed the "Troika" with Denmark and Mexico to facilitate "close cooperation" on the climate change negotiations (UNFCCC 2010). What is also interesting is the fact that at the Durban Summit, the South African Presidency introduced a negotiation tool or approach, called *Indaba*, that has now become a norm in the climate change negotiations. *Indaba*, "a Zulu tradition of people getting together to resolve a matter and not just re-stating their positions but trying to find a common ground and things which everyone can agree with" (IAF 2016), was used not only at the Durban Summit but also at the Paris Summit. This traditional tool has been useful in breaking deadlocks in the negotiations process so far, as negotiators gather around a table and speak to each other. *Indabas* symbolise "fairness" and "participatory", which are deemed to be at the core of South Africa's climate diplomacy by its leadership. The *Indaba* has been successful in cementing South Africa's position as a norm-giver. In any case, it was at the Durban Summit that the negotiations for the post-2020 agreement had been launched (Felix 2015). Therefore, it was in South Africa's special interest that the outcomes of the Durban Summit are carried through to fruition at the 2015 Paris Summit (Department of Environmental Affairs 2016; Zimmer 2015).

At the Paris Summit, South Africa – while vouching for development space for the developing countries, equity, and climate justice, as well as supporting the loss and damage mechanism established at the 2014 Warsaw Summit –[4] did not pose as strong a challenge to the developed countries' proposals on MRV (transparency) and legal instruments as China and India (Wang and Gao 2018). South Africa's low institutional capacity to deal with climate change and its increasing vulnerability to it in light of the recent droughts in the country have rendered a sense of urgency among the political class as well as the civil society. The "Day Zero" water crisis (when taps are switched off and water rationing begins) pertaining to the city of Cape Town brought attention to the need for reducing water demand, saving water, and addressing climate change (Winter 2018). Yet, under former President Zuma, although internationally, pro-climate voices were dominant, domestically, pro-economy and pro-coal policies overrode environmental and climate priorities. The deployment of renewable energy projects was delayed too, due to pressure from entities such as Eskom – South Africa's state-owned grid operator as well as owner of South African coal-powered plants – thereby also reducing investments in the country's renewable energy sector (Climate Action Tracker n.d.).

Nevertheless, under the Paris Agreement, South Africa has pledged to peak its emissions sometime between 2020 and 2025 (Rich, Northrop and Mogelgaard 2015). In effect, South Africa is considered an example of a country whose climate diplomacy positions may not correspond with its domestic positions. Its strategy has been to tie climate diplomacy with the country's foreign policy agenda and delink it from the domestic climate action to a large extent, which is perhaps a pattern that could be associated with the other emerging economies as well. As pointed out by Ariane Labat,

> The emerging economies use the principle of 'equity' at the international level, but tend to use the rhetoric of 'competitiveness' at the domestic level. Most often these countries are looking for economic interests in climate solutions in order to be able to justify back home, whether and how they are good for their respective countries.

She also asserts that BASIC is united in terms of demanding certain commitments from the developed countries, but not so united when it comes to mitigation and adaptation solutions.[5]

South Africa's climate change commitments after the Paris Summit are driven to a large extent by socio-economic considerations. As already mentioned earlier, the idea of "just transition" is at the centre of its climate action, with sustainable development and poverty eradication continuing to dominate the discourse. The need for studying the "effects of decarbonisation on employment and the coal value chain" and putting in action "just transition" policies have been reiterated in several policy documents (Burton, Marquard and McCall 2019). President Cyril Ramaphosa, who took charge in 2018, continues to re-emphasise these aspects. In an address to the African Union (AU), not only did he reiterate the unethicality of expecting "Africa to shoulder burden of climate change alone", but he also expressed his regret at the manner in which many financiers at the international level are disinvesting from the fossil fuels industry hastily, without assessing its effects on jobs and economies (Hunter 2020). South Africa has often criticised the rich countries for providing "conditional loans" and "credit lines" instead of grants for undertaking climate action, which renders long-term and transformative initiatives difficult to implement

(Afedraru et al. 2018). Yet, under successive leaderships and with the level of regional integration in Africa, South Africa has been able to build institutional capacities, although the lack of resources and capacity as well as conflicts of interests between various domestic actors remain major hurdles.

South Africa also has the distinction of being one of the few developing countries to have introduced carbon tax. It has introduced not only an "electricity levy" but also a "CO_2 emissions levy on new motor vehicles imported into or manufactured in South Africa" (Steenkamp and Naude 2018). The carbon tax proposal has several undercurrents of poverty reduction discourse too. Some (mostly research, political/executive, and civil society actors) favoured it for its ability to contribute to poverty reduction measures. Yet, the poverty reduction narratives have been used by the businesses that are affected by the proposal to be more critical of it in terms of its effects on economic growth, jobs, and so on (Rennkamp 2019). Even though some experts allege that the tax is "cheap" and too "complicated", it is nevertheless a stepping-stone to achieving its Nationally Determined Contributions (NDCs) (Oxford 2019). These NDCs, according to President Ramaphosa, recognise South Africa as a "good global citizen", ready to contribute to the global effort to combat climate change, furthering the narrative on "responsibility", used by other emerging economies too. However, as is made clear in this section, the country's leadership has tied the global obligations to "national circumstances" and "development context" that should "inform" national climate policy, in its view. The developing country identity takes precedence over the "global citizen" identity, wherein the idea of "leaving no one behind" is underscored by the South African establishment (SA News 2019).

Notes

1 The author's interview with Matías Franchini, Assistant Professor of International Relations at the Universidad del Rosario, Bogotá via Skype on May 31, 2019.
2 The author's interview with Matías Franchini.
3 Based on the views expressed by Ana Flávia Barros-Platiau, University of Brasília, in a session on "South America in climate, biodiversity, forests, rivers, ocean and Antarctica governance: What are the drivers of poor performance?", at the Earth System Governance conference, hosted by the National Autonomous University of Mexico (UNAM), on November 6, 2019. The full programme of the conference is available here: https://www.earthsystemgovernance.org/mexico2019/programme/conference-schedule/.

4 The Warsaw International Mechanism for Loss and Damage associated with Climate Change Impacts was established "to address loss and damage associated with impacts of climate change, including extreme events and slow onset events, in developing countries that are particularly vulnerable to the adverse effects of climate change." More information is available on the website of the UNFCCC: https://unfccc.int/topics/adaptation-and-resilience/workstreams/loss-and-damage-ld/warsaw-international-mechanism-for-loss-and-damage-associated-with-climate-change-impacts-wim.

5 The author's interview with Ariane Labat, a member of the EU climate change negotiations delegation, via Skype on January 25, 2019.

Bibliography

Afedraru, Lominda, Waruru, Maina, Obi, Leopold, Makoni, Munyaradzi, Mhango, George, Christine, Chisha and Saujanya Shrivastav. 2018. "Funds to Fight Climate Change in Africa Grossly Inadequate." *Down to Earth*. June 3. Accessed July 4, 2020. https://www.downtoearth.org.in/news/climate-change/funds-to-fight-climate-change-in-africa-grossly-inadequate-60718.

AFP. 2019. "Merkel Wants 'Straight Talk' with Bolsonaro over Deforestation." *Space Daily*. June 26. Accessed July 4, 2020. https://www.spacedaily.com/afp/190626140722.4340bpio.html.

African Union Development Agency. n.d. *New Partnership for Africa's Development*. Accessed July 5, 2020. https://www.nepad.org/.

Amusan, Lere and Olutola, Oluwole. 2016. "Paris Agreement (PA) on Climate Change and South Africa's Coal-Energy Complex: Issues at Stake." *Africa Review* 9 (1): 43–57.

Belis, David, Schunz, Simon, Wang, Tao and Jayaram, Dhanasree. 2018. "Climate Diplomacy and the Rise of 'Multiple Bilateralism' between China, India and the EU." *Carbon & Climate Law Review* 12 (2): 85–97.

Boadle, Anthony. 2014. "Brazil's Silva Launches Bid, Threatens Rousseff Re-Election." *Reuters*. August 21. Accessed July 4, 2020. https://www.reuters.com/article/us-brazil-election/brazils-silva-launches-bid-threatens-rousseff-re-election-idUSKBN0GK2HS20140821.

Burton, Jesse, Marquard, Andrew and McCall, Bryce. 2019. "Socio-Economic Considerations for a Paris Agreement-Compatible Coal Transition in South Africa." *Climate Transparency Policy Paper*. July. Accessed July 4, 2020. https://www.climate-transparency.org/wp-content/uploads/2019/07/CT-Just-Transition-in-South-Africa.pdf.

Climate Action Tracker. n.d. "Country Summary: South Africa." *Climate Action Tracker*. Accessed July 4, 2020. https://climateactiontracker.org/countries/south-africa/.

Department of Environmental Affairs. 2016. "South Africa Welcomes the Adoption of the Paris Agreement." *The New Age*. January 27. https://www.environment.gov.za/mediarelease/southafrica_welcomesadoption_parisagreement.

Doherty, Ben. 2019. "China Accuses Australia of Being a 'Condescending Master' in the Pacific." *The Guardian*. August 21. Accessed July 4, 2020. https://www.theguardian.com/australia-news/2019/aug/21/china-accuses-australia-of-being-a-condescending-master-in-the-pacific.

Edwards, Guy and Roberts, Timmons. 2015. "Latin America and UN Climate Talks: Not in Harmony." *Americas Quarterly*. February 3. Accessed July 4, 2020. https://www.americasquarterly.org/fulltextarticle/latin-america-and-un-climate-talks-not-in-harmony/.

Escobar, Herton. 2019. "Brazil's New President Has Scientists Worried. Here's Why." *American Association for the Advancement of Science*. June 22. Accessed July 4, 2020. https://www.sciencemag.org/news/2019/01/brazil-s-new-president-has-scientists-worried-here-s-why.

Evans, Simon and Gabbatiss, Josh. 2019. "In-Depth Q&A: How 'Article 6' Carbon Markets Could 'Make or Break' The Paris Agreement." *Carbon Brief*. November 29. Accessed July 5, 2020. https://www.carbonbrief.org/in-depth-q-and-a-how-article-6-carbon-markets-could-make-or-break-the-paris-agreement.

Farias, Déborah B. L. 2015. "Brazil's 'Ethanol Diplomacy': Technical Cooperation as a Foreign Policy Tool." *Conference Paper*. New Orleans: International Studies Association.

Felix, Bate. 2015. "Climate Talks Turn to South African Indaba Process to Unlock Deal." *Reuters*. December 10. Accessed July 4, 2020. https://www.reuters.com/article/us-climatechange-summit-indaba/climate-talks-turn-to-south-african-indaba-process-to-unlock-deal-idUSKBN0TT29C20151210.

Gallagher, Kelly Sims and Xuan, Xiaowei. 2019. *Titans of the Climate: Explaining Policy Process in the United States and China*. Cambridge: MIT Press.

Grudgings, Stuart. 2008. "Brazil's President Puts Economy Ahead of Amazon." *Reuters*. May 15. Accessed July 4, 2020. https://www.reuters.com/article/us-brazil-environment-idUSN1451777720080514.

Hallding, Karl, Olsson, Marie, Atteridge, Aaron, Vihma, Antto, Carson, Marcus, and Román, Mikael. 2011. "Together Alone: BASIC Countries and the Climate Change Conundrum." *SEI*. November 28. Accessed August 7, 2020. https://www.sei.org/publications/together-alone-basic-countries-and-the-climate-change-conundrum/.

Hochstetler, Kathryn Ann. 2012. "The G-77, BASIC, and Global Climate Governance: A New Era in Multilateral Environmental Negotiations." *Revista Brasileira de Política Internacional* 55. Accessed July 20, 2020. http://dx.doi.org/10.1590/S0034-73292012000300004.

Hung, Ming-Te, and Tsai, Tung-Chieh. 2012. "Dilemma of Choice: China's Response to Climate Change." *Revista Brasileira de Política Internacional* 55: 104–124.

Hunter, Qaanitah. 2020. "Unfair to Expect Africa to Shoulder Burden of Climate Change Alone, Ramaphosa Tells AU." *Times Live*. February 8. Accessed July 5, 2020. https://www.timeslive.co.za/politics/2020-02-08-unfair-to-expect-africa-to-shoulder-burden-of-climate-change-alone-ramaphosa-tells-au/.

IAF. 2016. "Traditional Meeting Technique Powers Climate Change Breakthrough." *International Association of Facilitators.* December. Accessed July 4, 2020. https://www.iaf-world.org/site/global-flipchart/6/indaba.

Jaeger, Joel, Joffe, Paul and Song, Ranping. 2017. "China Is Leaving the U.S. Behind on Clean Energy Investment." *World Resources Institute.* January 6. Accessed July 3, 2020. https://www.wri.org/blog/2017/01/china-leaving-us-behind-clean-energy-investment.

Jiahan, Cao. 2018. "Recalibrating China-U.S. Climate Cooperation under the Trump Administration." *China Quarterly of International Strategic Studies* 4 (1): 77–93.

Jing, Li. 2015. "A Climate for Change: How China Went from Zero to Hero in Fight against Global Warming in Just 6 Years." *South China Morning Post.* November 27. Accessed July 4, 2020. https://www.scmp.com/news/china/policies-politics/article/1884037/climate-change-how-china-went-zero-hero-fight-against.

Kaneti, Marina. 2020. "China's Climate Diplomacy 2.0." *The Diplomat.* January 2. Accessed July 3, 2020. https://thediplomat.com/2020/01/chinas-climate-diplomacy-2-0/.

Kiessling, Christopher Kurt. 2018. "Brazil, Foreign Policy and Climate Change (1992–2005)." *Contexto Internacional* 40 (2): 387–407.

Kotzé, Louis Jacobus, Humby, Tracy, Rumble, Olivia and Gilder, Andrew. 2016. *Climate Change Law & Governance in South Africa.* Cape Town: Juta.

Landler, Mark. 2014. "U.S. and China Reach Climate Accord after Months of Talks." *New York Times.* November 11. Accessed July 4, 2020. https://www.nytimes.com/2014/11/12/world/asia/china-us-xi-obama-apec.html?_r=0.

Lehr, Deborah. 2019. "Is China Still the Global Leader on Climate Change?" *The Diplomat.* October 21. Accessed July 4, 2020. https://thediplomat.com/2019/10/is-china-still-the-global-leader-on-climate-change/.

Li, Anthony H. F. 2016. "Hopes of Limiting Global Warming? China and the Paris Agreement on Climate Change." *China Perspectives* 1: 49–54.

Li, Lina and Ivleva, Daria. 2019. "China as a New Climate-Responsible Donor?" *Climate Diplomacy.* January 28. Accessed July 4, 2020. https://www.climate-diplomacy.org/publications/china-new-climate-responsible-donor.

Liu, Coco. 2017. "The Real Reason for China's U-turn on Climate Change." *South China Morning Post.* February 4. Accessed July 4, 2020. https://www.scmp.com/week-asia/geopolitics/article/2067189/real-reason-chinas-u-turn-climate-change.

Luft, Gal. 2013. "China's Pollution Revolution." *Foreign Policy.* December 16. Accessed July 3, 2020. https://foreignpolicy.com/2013/12/16/chinas-pollution-revolution/.

Ministério do Meio Ambiente. 2009. Dilma and Minc Announce That Brazil Can Donate US$ 5 Billion to Poor Countries." *Ministry of Environment of Brazil.* December 16. Accessed July 4, 2020. https://www.mma.gov.br/informma/item/6015-dilma-and-minc-announce-that-brazil-can-donate-us-5-billion-to-poor-countries.

Nature. 2018. "Brazil's New President Adds to Global Threat to Science." *Nature* 563: 5–6. Accessed July 4, 2020. https://www.nature.com/articles/d41586-018-07236-w.

News Wires. 2019. "Macron Spearheads Pressure on Bolsonaro over Amazon Fires." *France 24*. August 24. Accessed July 4, 2020. https://www.france24.com/en/20190824-macron-france-brazil-bolsonaro-amazon-fires.

Nunez, Christina. 2017. "China Poised for Leadership on Climate Change after U.S. Reversal." *National Geographic*. March 28. Accessed July 4, 2020. https://www.nationalgeographic.com/news/2017/03/china-takes-leadership-climate-change-trump-clean-power-plan-paris-agreement/.

Oxford, Tasmin. 2019. "The Carbon Tax: Why It Could Work." *Mail and Guardian*. August 23. Accessed July 4, 2020. https://mg.co.za/article/2019-08-23-00-the-carbon-tax-why-it-could-work/.

Pashley, Alex. 2015. "Brazil Backs 'High Ambition Coalition' to Break Paris Deadlock." *Climate Home News*. December 11. Accessed July 4, 2020. https://www.climatechangenews.com/2015/12/11/brazil-backs-high-ambition-coalition-to-break-paris-deadlock/.

Pashley, Alex. 2015. "Brazil: Redeemer of a Paris Climate Deal?" *Climate Home News*. November 18. Accessed July 4, 2020. https://www.climatechangenews.com/2015/11/18/brazil-redeemer-of-a-paris-climate-deal/.

Paskal, Cleo. 2020. "How Kiribati Was Lost to China." *The Sunday Guardian*. August 29. Accessed August 29, 2020. https://www.sundayguardianlive.com/news/kiribati-lost-china.

PIB. 2019. "Joint Statement Issued at the Conclusion of 29th BASIC Ministerial Meet on Climate Change." *Press Information Bureau, Government of India*. October 26. Accessed July 4, 2020. https://pib.gov.in/Pressreleaseshare.aspx?PRID=1589318.

PTI. 2018. "Donald Trump Blames India, China for His Decision to Withdraw from Paris Climate Deal." *Economic Times*. February 24. Accessed July 4, 2020. https://economictimes.indiatimes.com/news/international/world-news/donald-trump-blames-india-china-for-his-decision-to-withdraw-from-paris-climate-deal/articleshow/63057148.cms.

Rennkamp, Britta. 2019. "Power, Coalitions and Institutional Change in South African Climate Policy." *South African Journal of International Affairs* 19 (6): 756–770.

Rennkamp, Britta and Marquard, Andrew. 2018. "South Africa's Multiple Faces in Current Climate Clubs." *South African Journal of International Affairs* 24 (4):443–461.

Reuters. 2015. "Norway to Complete $1 Billion Payment to Brazil for Protecting Amazon." *Reuters*. September 15. Accessed July 4, 2020. https://www.reuters.com/article/us-climatechange-amazon-norway-idUSKCN0RF1P520150915.

Reuters. 2015. "Obama Calls Xi to Praise China's Role in Climate Talks – W. House." *Reuters*. December 14. Accessed July 4, 2020. https://af.reuters.com/article/commoditiesNews/idAFL1N1430ZV20151214.

Reuters. 2019. "Igniting Global Outrage, Brazil's Bolsonaro Baselessly Blames NGOs for Amazon Fires." *CNBC*. August 22. Accessed July 4, 2020. https://www.cnbc.com/2019/08/22/brazil-jair-bolsonaro-baselessly-blames-ngos-for-amazon-fires.html.

Rich, David, Northrop, Eliza and Mogelgaard, Kathleen. 2015. "South Africa Pledges to Peak Its Greenhouse Gas Emissions by 2025." *World Resources Institute*. October 1. Accessed July 5, 2020. https://www.wri.org/blog/2015/10/south-africa-pledges-peak-its-greenhouse-gas-emissions-2025.

SA News. 2019. "South Africa Committed to Climate Change Interventions." *South Africa Government News Agency*. September 24. Accessed July 4, 2020. https://www.sanews.gov.za/south-africa/south-africa-committed-climate-change-interventions.

Schmidt, Jake. 2010. "The Copenhagen Accord: Foundations for International Action on Climate Change." *Natural Resources Defense Council*. June 23. Accessed July 3, 2020. https://www.nrdc.org/resources/copenhagen-accord-foundations-international-action-climate-change.

Schoenwalder, Cecelia Smith. 2019. "The Amazon Rainforest Is on Fire. Can Pressure from Other Nations Get Brazil to Act?" *US News*. August 23. Accessed July 4, 2020. https://www.usnews.com/news/world-report/articles/2019-08-23/the-amazon-rainforest-is-on-fire-can-pressure-from-other-nations-get-brazil-to-act.

SCMP Reporters. 2020. "What Is the US-China Trade War?" *South China Morning Post*. April 13. Accessed July 4, 2020. https://www.scmp.com/economy/china-economy/article/3078745/what-us-china-trade-war-how-it-started-and-what-inside-phase.

Sethi, Nitin. 2010. "'We Can't Shut Door on Talks'." *Economic Times*. December 10. Accessed July 4, 2020. https://economictimes.indiatimes.com/news/environment/global-warming/we-cant-shut-door-on-talks/articleshow/7074386.cms?from=mdr.

Sheng, Yang. 2019. "Xi Article Shows Determination to Build Ecological Civilization." *Global Times*. February 1. Accessed July 3, 2020. http://www.globaltimes.cn/content/1137873.shtml.

Singh, Mandip. 2013. "Malacca: No More a Dilemma for China?" *Centre for Land Warfare Studies – Scholar Warrior*. Accessed July 4, 2020. https://archive.claws.in/journal/journal-scholar-warrior/index.html.

Smith, Alexander. 2017. "China's Xi Lectures Trump on Globalization and Climate Change." *NBC News*. January 17. Accessed July 4, 2020. https://www.nbcnews.com/news/world/china-s-xi-lectures-trump-globalization-climate-change-n707721.

Stats SA. n.d. "The Importance of Coal." *Statistics Department South Africa*. Accessed July 4, 2020. http://www.statssa.gov.za/?p=4820.

Steenkamp, Lee-Ann and Naude, Piet. 2018. "Some Ethical Considerations for South Africa's Climate Change Mitigation Approach in Light of the Paris Agreement." *African Journal of Business Ethics* 12 (2): 70–84.

Sugiuira, Eri and Okutsu, Akane. 2018. "China's Renewable Energy Surges after State Backing." *Nikkei Asian Review.* November 21. Accessed July 4, 2020. https://asia.nikkei.com/Spotlight/The-Big-Story/China-s-renewable-energy-surges-after-state-backing.

Tabau, Anne-Sophie and Lemoine, Marion. 2012. "Willing Power, Fearing Responsibilities: BASIC in the Climate Negotiations." *Carbon & Climate Law Review* 6 (3): 197–208.

The Economist. 2013. "The East Is Grey." *The Economist.* August 10. Accessed July 4, 2020. https://www.economist.com/briefing/2013/08/10/the-east-is-grey.

The White House. 2015. "U.S.-Brazil Joint Statement on Climate Change." *The White House.* June 30. Accessed July 4, 2020. https://obamawhitehouse.archives.gov/the-press-office/2015/06/30/us-brazil-joint-statement-climate-change.

Timperley, Jocelyn. 2018. "The Carbon Brief Profile: Brazil." *Carbon Brief.* March 7. Accessed July 4, 2020. https://www.carbonbrief.org/the-carbon-brief-profile-brazil.

Timperley, Jocelyn. 2019. "Brazil Fights Attempt to Cancel Its Old Carbon Credits." *Climate Home News.* 11 October. Accessed July 4, 2020. https://www.climatechangenews.com/2019/10/11/brazil-fights-attempt-cancel-old-carbon-credits/.

UNFCCC. 2010. "Troika Meeting of Host Countries of COP 15, COP 16 and COP 17." *United Nations Framework Convention on Climate Change.* September 24. Accessed July 5, 2020. https://unfccc.int/files/meetings/cop_16/application/pdf/100924_troika_declaration_new_york.pdf.

UNFCCC. 2015. "South Africa's Intended Nationally Determined Contribution (INDC)." United Nations Framework Convention on Climate Change. Accessed July 4, 2020. https://www4.unfccc.int/sites/ndcstaging/Published Documents/South%20Africa%20First/South%20Africa.pdf.

UNSTATS. 2008. "South Africa Energy Statistics." *United Nations.* Accessed August 10, 2020. https://unstats.un.org/unsd/energy/meetings/mexico2008/Country%20notes/South%20Africa.pdf.

Viola, Eduardo and Franchini, Matías. 2014. "Brazilian Climate Politics 2005–2012: Ambivalence and Paradox." *Climate Change* 5 (5): 677–688.

Viola, Eduardo and Franchini, Matías. 2017. *Brazil and Climate Change.* New York: Routledge.

Wang, Tian and Gao, Xiang. 2018. "Reflection and Operationalization of the Common but Differentiated Responsibilities and Respective Capabilities Principle in the Transparency Framework under the International Climate Change Regime." *Advances in Climate Change Research* 9 (4): 253–263.

Watts, Joshua and Depledge, Joanna. 2018. "Latin America in the Climate Change Negotiations: Exploring the AILAC and ALBA Coalitions." *Wires: Climate Change* 9 (6): e533.

Winter, Kevin. 2018. "Day Zero Is Meant to Cut Cape Town's Water Use: What Is It, and Is It Working?" *The Conversation.* February 20. Accessed July 4, 2020. https://theconversation.com/day-zero-is-meant-to-cut-cape-towns-water-use-what-is-it-and-is-it-working-92055.

World Bank. 2018. "South Africa Economic Update." *World Bank*. April. Accessed July 5, 2020. http://pubdocs.worldbank.org/en/798731523331698204/South-Africa-Economic-Update-April-2018.pdf.

Xiaosheng, Gao. 2018. "China's Evolving Image in International Climate Negotiation." *China Quarterly of International Strategic Studies* 4 (2): 213–239.

Yeo, Sophie. 2015. "Analysis: Brazil's Climate Pledge Represents Slight Increase on Current Emissions." *Carbon Brief*. September 29. Accessed July 4, 2020. https://www.carbonbrief.org/analysis-brazils-climate-pledge-represents-slight-increase-on-current-emissions.

Zimmer, Ben. 2015. "The African Discussion Style 'Indaba' Thrived at Climate Talks." *The Wall Street Journal*. December 18. Accessed July 4, 2020. https://www.wsj.com/articles/the-african-discussion-style-indaba-thrived-at-climate-talks-1450456996.

4 The evolution of India's climate diplomacy (2009–2013)

Existing discourse on India's ideational positions in the negotiations

India's climate diplomacy positions can essentially be traced back to the speech delivered by former Prime Minister Indira Gandhi at the 1972 UNHCE (as explicated in the introductory chapter), wherein she drew correlations between environment, development, and poverty. Since then, India has steadfastly maintained continuity in its climate diplomacy positions, focussed upon a formula comprising the following ideas, values, and principles: historical responsibility, development space, strategic autonomy, inalienable right to utilise its resources to eradicate poverty, non-interference in the internal affairs (or sovereignty) among others (Dasgupta 2019; Sengupta 2019). Dubash (2009) seeks to explain India's positions in the global climate change negotiations through the lens of domestic politics, by categorising the positions into three, as explained in Table 4.1.

The framing of climate change has been dominated by "historical and developmental" concerns, with the thrust on the developed countries' responsibility towards acting on it and the developing countries' priorities towards poverty reduction (MEA 2008). Even in the Indian media, despite increased focus on India's climate vulnerabilities, this framing is dominant. For instance, Billett (2009) regards the Indian media's coverage of climate change as a way of labelling it as "international carbon colonialism" and concomitantly articulating "a nationalistic narrative". When asked about the international community's responsibility towards acting on climate change, for instance, Nitin Sethi, an environmental journalist, opines that India's argument that "countries must act based on both their historical responsibilities and capabilities and not just the latter" is "scientific, ethical as well as strategically rock solid". He adds, "What a nation needs to think is what

Table 4.1 Narratives on India's Climate Diplomacy Positions

Group	Perceptions about Climate Change and Diplomacy
Growth-first stonewallers	• Not entirely convinced by the challenges posed by climate change to India. • Unwilling to compromise on the growth agenda. • See the climate change negotiations as a Western agenda to prevent countries such as India from developing. • Focus on "equity across nations" – making their perception of equity more strategic than principled.
Progressive realists	• Acknowledging the impacts of climate change on India. • "Cynical" about the climate change negotiations for the lack of consideration given to historical responsibility and equity. • Criticise the developed world's strategy of hiding behind countries such as India and China to not act upon a problem, primarily created by it. • Uphold the per capita principle as a legitimate foundation on which the international climate regime could be built upon. • Domestically, push for a low-carbon growth strategy, seeing the co-benefits of climate action, thus pursuing "environmental sustainability" and "internal equity".
Progressive internationalists	• Acknowledge climate vulnerabilities and effects on the global poor. • More favourable towards adopting international commitments, opting for both "equity" and "effectiveness" in an increasingly climate-vulnerable world, and in order to be able to influence the negotiations dynamics. • Link domestic actions with the climate regime, which would also put pressure on the developed countries to commit and act more. • Advance the formula of "seizing the moment" and make India a "first-mover in developing low-carbon technologies", which is considered to be economically beneficial for the country.

are its interests. And if our national interests and priorities have not changed why should our stance change?" (Sethi 2015). Similarly, Sunita Narain, Director of the Centre for Science and Environment (CSE), an independent think tank based in New Delhi, categorically states, "The issue of climate change is connected to growth and development.

No country in the world can accept a freeze on its development. Countries that have overused their share of global atmospheric space have a natural debt to pay" (VOA News 2010).

India's framing of climate change as a problem of equity more than the environment, and as a diplomatic challenge rather than a developmental one for a long time (Dubash 2009, pp. 14–15), had put it on the defensive in the international climate change negotiations. Its inability to take into account environmental considerations associated with climate change sufficiently could be considered the reason for its disregard of the need for it to take on emissions reduction commitments in the future. Moreover, it also created a vacuum as far as the domestic pool of scientific and technical knowledge on climate change is concerned. At the domestic level, the National Action Plan on Climate Change (NAPCC) was launched in 2008 and several other regulations were also introduced that could be linked to climate change. This was an indication of India's willingness to change this formula and tie climate goals with development priorities, as long as the latter would not be hindered. Nevertheless, it delinked its domestic actions from international commitments. Shyam Saran pointed out that

> whatever action we [India] take domestically to pursue sustainable development, let it be clearly understood that there is no legal obligation on the part of India, under existing international instruments, to take on binding emissions reduction obligations, now or in the post 2012 period.
>
> (Saran 2008)

This led to India being labelled "obdurate", "intransigent", "renegade" etc. (Ramesh 2009b), which may be considered unfair. According to the "Singh principle", named after Prime Minister Manmohan Singh, India's per capita emissions are comparatively far lower than that of all the developed countries and even most developing countries, and they would never surpass that of the developed world even in the future. This pledge was declared by him while releasing the NAPCC (Ghosh 2008). Moreover, India's actions were and are always contingent on the actions of the developed countries – specifically, their commitment to reduce emissions by 40 percent by 2020, as per 1990 levels (under the Kyoto Protocol) (Ramesh 2009b). This in itself was a nuanced shift in India's climate diplomacy. Not only did India begin to put in place domestic measures to tackle climate change, but it also realised the importance of being considered a part of the solution, in light of the tags it received in the negotiations and even international media. In short, the Copenhagen Summit and the COPs that came thereafter saw a gradual increase in the influence of "progressive internationalists" within India's climate diplomacy discourse.

In their work, Saran and Jones (2017, pp. 66–88) explore varying climate identities of India. Among the factors, one of the most influential is the "rural identity", considering that India is a largely agricultural country, with more than 60 percent of the people still living in rural areas (Trading Economics n.d.). In addition, it builds on the perception that India's approach to climate change should be adaptation-centric. The second identity is that of "energy security identity" (Saran and Jones 2017), which has become a cornerstone of India's climate policy, both domestically and internationally. Though India's coal dependence is expected to continue in the foreseeable future, this identity is pushing the administration to expand the scope of renewable and other forms of clean energy in the country's energy basket, thereby promoting energy self-reliance or self-sufficiency, taking into account its excessive dependence on imported fossil fuels. Third, India's industrial identity is becoming more and more dominant, particularly with the Modi administration's launch of "Make in India" (which is discussed further in Chapter 5). Since India is relatively late in the race for industrialisation, it is also mindful of the consequences of expansion of its manufacturing base to meet its developmental requirements, thus injecting the idea of "green economy" into the industrialisation narrative.

Fourth, India's "entrepreneurial identity" is, in a way, related to its industrial identity, focussing on "green market, manufacturing sector and industrial efficiency" (Saran and Jones 2017). The fifth identity is that of a "developing nation", one that draws upon its membership of the G-77, steeped in the principles/ideas of CBDR-RC, historical responsibility, and right to development. Last but not the least is India's "emerging nation identity", in line with its partnership with the other BASIC or BRICS countries. It is through this identity that India endeavours to project itself as a leader in the climate change negotiations, as well as that of the developing world. Interestingly, according to the authors, this identity also highlights India's ambivalence in its climate diplomacy as it seeks to take the middle ground, wherein it continues to uphold the principle of CBDR-RC. At the same time, it attempts to "manage its interests with other developing as well as developed countries", and at times in favour of a legal instrument to bind India's commitments – to be treated as a "responsible stakeholder" internationally (Saran and Jones 2017). As the authors gather through their interviews, they find that the respondents are divided on the impact of domestic drivers of India's international positions, with some raising issues of poverty reduction and others (especially young professionals) disconnecting foreign policy objectives from the domestic ones (Saran and Jones 2017). It is important to note that at different points in time, different identities have been dominant, as would be clear by the end of this chapter and Chapter 5, but at no point has a single identity been completely erased, which makes India's case a peculiar one – that of reconciliation of diametrically opposite identities that are directly linked to its climate diplomacy positions.

In the next section, India's climate diplomacy under the Singh administration is looked into. There are several reasons why the two regimes have been analysed separately. First, the shift from India's desire to be recognised as a responsible power to being exalted as a climate leader, albeit subtle, took place during the transition from the Singh administration to Modi administration. However, it should also be noted that the Modi administration's climate diplomacy positions were merely a continuation of that of the Singh administration – building on the steps taken by the latter since the Copenhagen Summit. Second, the influence of certain individuals in the Singh administration such as Shyam Saran and Jairam Ramesh on India's climate diplomacy decision-making was evident. On the contrary, under PM Modi, climate diplomacy has been further centralised and, perhaps, personalised with his ideas of "multilateralism" and *Panchamrit* (that is covered in Chapter 5). Third, at the international level too, many changes were witnessed since 2014, with the momentum for the Paris Summit increasing and many geopolitical realignments emerging (such as the US–China joint communiqué).

India's climate diplomacy under the Manmohan Singh administration with a focus on 2009–2013

Moral/ethical ideas

Among the moral or ethical ideas that India has espoused since the beginning of the climate change negotiations are equity and climate justice, which have guided principles of CBDR-RC, "historical responsibility", per capita emissions, "the polluter pays" etc. This section examines these concepts further from an Indian perspective and analyses how they have shaped India's climate diplomacy positions over a period of time. In the run-up to the Copenhagen Summit, India insisted on the continuation of the Kyoto Protocol and the enforcement of its second commitment period, based on the principle of CBDR-RC. According to India, this principle should be an inalienable moral or ethical foundation of any agreement reached within the UNFCCC. However, it was clear that the majority of developed countries were not in favour of this move, unless countries such as China and India also came on board the arrangement. Nevertheless, Jairam Ramesh's show of support for "binding commitments under appropriate legal form" and the replacement of "equal per capita entitlement principle" with "equitable access to sustainable development" at the 2010 Cancun Summit (Mehta 2011) is considered one of the major shifts in India's climate diplomacy positions, much to the furore of the other domestic constituents within India (Hilton 2010). With this slight shift in stance, India's positions on equity and international commitments took a different turn, albeit not drastically divergent from its previously held positions.

Importantly, even while agreeing to these changes, India pushed for differentiation between the developed and developing countries, which has been at the centre of the negotiations since the Cancun Summit, up until the Paris Agreement was signed. As Navroz Dubash comments, "India's push for differentiation began to be more influenced by the need for its operationalisation rather than being treated as a principle".[1] However, Chandrasekhar Dasgupta, who has been a part of India's negotiating team for a long time, criticised Ramesh for undermining India's position on "verification" and "equity", and acceding to commitments under an "appropriate legal forum" (Dasgupta 2014). Ramesh's stance was based on the argument that "equitable access to carbon space" somehow connoted "right to pollute". Moreover, he brought up another point that had begun to tweak India's stance on equity, in light of the growing inequality at the domestic level, which according to him can only be addressed through equitable access to sustainable development (Ramesh 2015). According to figures, there is a significant rural–urban disparity in emissions, with the average urban emissions being about 2.5 times the rural emissions and the inequality higher when it comes to emissions in urban India itself (Chakravarty and Ramana 2011). In such a scenario, India was being accused of hiding behind its poor and not being accountable for its rising emissions.

One could, therefore, claim that a change in mood was steered by personalities, state institutions, and other relevant actors (highlighted in the subsequent sections), which led to the creation of a new formula, albeit with resistance from several quarters. On the one hand, the developed countries' promise of providing additional finance and technology to developing countries since 1992 remained unfulfilled, in addition to the US's non-ratification of the Kyoto Protocol and several other developed countries' (including Canada, Australia, and Japan) unenthusiastic response during the first commitment of the Kyoto Protocol. On the other, the strict differentiation between the developed and developing countries as well as the equity principle were being diluted gradually by the developed countries, which India eventually began to compromise upon.

Jairam Ramesh's new formula of "per capita plus", which is to stick to the per capita principle but also to offer more than just that, had begun to take shape in more practical terms (Ghosh 2009). In a leaked letter to the Indian press before the Copenhagen Summit, his suggestions to the Prime Minister included "domestic law for climate change mitigation", "welcoming the US into the mainstream", and the need for India "to be seen as more pragmatic and constructive" (Vihma 2011). Ramesh believed that a "change in stance" in the climate change negotiations, including by allowing international scrutiny of Indian climate mitigation actions, could even give India a permanent seat in the UNSC (Sethi 2009). However, due to widespread criticism, they could not be implemented in entirety. He also appointed an economist, Arvind Subramanian (who went on to take charge as the Chief Economic Adviser to the Government of India in 2014), to define equity. His views on the existing operational definitions of equity have been expressed in a co-authored article (Birdsall et al. 2009), in which he blames the "limited progress in climate change discussions" to the "arbitrary" nature of "burden-sharing" proposals that essentially are hinged on the "wrong starting point" of "the right to pollute", thereby widening the gap between the "rich and poor worlds".

Even though the idea of equity is considered to have taken the backseat during 2009–2011 due to geopolitical and other shifts (as mentioned below), it was not entirely shunned by the Indian establishment. With the exit of Jairam Ramesh from the ministry, India's stance had been focussed more on ensuring that the second commitment period of the Kyoto Protocol came into effect and that equity was deeply embedded in the future agreement, in the context of it being "applicable to all", as had already been agreed upon. At the 2011 Durban Summit, when the parties agreed to put the phrase – "an agreed outcome with legal force" – in the final outcome, this was a major diplomatic victory for India, which was opposed to the use of "protocol" and "legal instrument" in it. India's Minister for Environment and Forests, Jayanthi Natarajan (IANS 2011), even while acceding to several demands, brought back to the fore India's "red lines" on "livelihoods" and "equity" among others.

With this, India's stance had somewhat hardened as the fear of further dilution of differentiation between the developed and developing countries became more imminent. The statements by Natarajan also brought back the colonial/neo-colonial frame that India often used in the 1990s and 2000s to oppose the rich countries' attempts to impose their agenda on the rest of the world, particularly developing countries. At the subsequent Doha (2012), Warsaw (2013), and Lima (2014) Summits, India pushed fervently for the inclusion of equity and CBDR-RC in the post-2020 agreement. The LMDC came together before the Doha Summit. They asserted that the Ad Hoc Working Group on the Durban Platform for Enhanced Action (ADP)[2] negotiations "shall be guided by and must be consistent with the principles and provisions of the Convention, especially the principles of equity and common but differentiated responsibilities". More importantly, India anchored decisively to its core support group – the G-77 plus China – "to protect the overall interests of developing countries" (Lok Sabha Secretariat 2013).

Environmental/ecological ideas

One of the characteristics of India's positions on climate change in general is its growing consciousness of the vulnerabilities and threats posed by climate change to the country's economy and societies. This change in perceptions is said to have influenced India's willingness to prepare itself for any eventuality and focus on studying the impacts of climate change on the country. One could argue that the establishment of the National Action Plan on Climate Change (NAPCC) in 2008 (Figure 4.1) was a reflection of the increasing cognisance at the domestic level of the challenges posed by climate change to the country, and not so much a response to the evolving international dynamics. However, as experts contend, the NAPCC was not just in consonance with the requirement of NAMAs in light of the shifts in the climate change negotiations since the 2007 Bali Summit but also meant largely for "international consumption".[3] Nevertheless, the biggest factor that triggered this major shift was the impact of climate change on the southwest monsoon, on which India's GDP is dependent heavily. Since India's agricultural sector (largest source of livelihoods in India) is also heavily dependent on the monsoon (with more than 50 percent of the net sown area that is unirrigated) (Varma 2015), the impacts of rainfall variations on crop yields and livelihoods could be disastrous and, hence, had to be addressed.

The other two factors that featured prominently in the discussions preceding the release of the NAPCC and thereafter (in the run-up to the 2009 Copenhagen Summit) were the impacts of climate change on India's coastline (susceptible to sea level rise and extreme weather events) and the Himalayan glaciers that feed into the rivers of the Ganges, Indus and Brahmaputra basins, flowing through India, and thus critical for water security (Ramesh 2009a). When the action plan was released, the reactions from research institutes, environmental journalists, and environmentalists were mixed. While some hailed it as a step in the right direction with the possibility of India being different by "leapfrogging to a low carbon economy using high-end and emerging technologies", others called it a document that lacked vision, as it "put economic development ahead of emission reduction targets" (Pandve 2009). In any case, India's perceptions of its vulnerability had undergone a major shift, which began to influence its domestic outlook towards climate action and policy, which had been neglected for most of the 1990s and early 2000s.

Figure 4.1 India's National Action Plan on Climate Change.

In addition, the Indian establishment had become increasingly aware of the need for creating knowledge on the causes and impacts of climate change domestically. This is primarily in the wake of studies that exaggerated the pace at which the Himalayan glaciers were melting and the Indian agricultural sector's contribution to methane emissions (Ramesh 2017). In 2007, the Intergovernmental Panel on Climate Change (IPCC) made an erroneous assessment/claim that there were high chances of the Himalayan glaciers melting away by 2035, which the IPCC's officials, including its then chairman, R. K. Pachauri (an Indian himself), later admitted was "unfounded" (Carrington 2010), after a group of Indian glaciologists refuted the claim based on their scientific study.

In terms of the positioning itself at the international level, India had begun to question the use of science to drive political agendas, in conjunction with scientists acting as "evangelists" (Ramesh 2017). In short, the need for reducing dependence on West-driven scientific studies was felt and the Indian Network for Climate Change Assessment (INCCA) was launched in 2009 (comprising "125 research institutions spread across the country with about 250 scientists from various Ministries"). The INCCA went on to release the Climate Change Assessment for 2030 in 2010, with a focus on four sectors – agriculture, water, natural ecosystems and biodiversity, and health – and four climate sensitive regions – Himalayan region, the Western Ghats, the Coastal Area, and the North-East Region (MoEF 2010). With this, India had embarked on the path of expanding its capacity in the field of climate science, which also gave it more confidence in its climate diplomacy with the rest of the world (Dubash et al. 2018).

Cultural ideas

The cultural ideas that India has espoused in its climate diplomacy positions are mostly latent and are not explicitly stirred up from time to time. In any case, these ideas have become stronger with the evolution of time, especially during the Modi administration (explored further in Chapter 5). One of the most oft-quoted set of ideas in India's climate diplomacy are India's "strong environmental ethic", whether it is in terms of the choice of food and automobiles or recycling goods and waste (Narain et al. 2009). Based on this argument, it has been reiterated by Indian delegates in the climate change negotiations that India does not need to be tutored on climate action by the West and that it would "find its own way" in climate action (and that it does not need to follow the West's footsteps) (Billett 2009). India's world view when it comes to climate change and climate action is guided by its deeper acknowledgement of the harmonious coexistence between human beings and the "Nature's cycle of birth, growth, decay and regeneration" (Saran 2009), as reiterated by Shyam Saran in one of his speeches. He continues:

> India draws upon its civilizational legacy in raising public awareness and promote community activism and initiative on Climate Change. Safeguarding the environment, looking upon Nature, not as a dark force to be conquered and subdued, but as a Mother, and a source of nurture, to be respected and preserved, is a concept deeply ingrained in Indian tradition.

It is often claimed by the Indian policy-makers that sustainable development, conservation, and resource management have been a part and parcel of the Indian culture, especially in the pre-colonial era. These claims bring to fore the prevalence of nature worship in India since ancient times. For example, the Bishnoi tribe of Rajasthan in India has made ecological sensibility and protection of flora and fauna the cornerstones of their religion (Bikku 2019). In addition, the cultural aspects interact with the perceptions of climate change as far as vulnerabilities are concerned. For instance, rivers such as the Ganges are extremely important for millions of Indians, not just from a socio-economic perspective but also from a cultural perspective. When these rivers and other such ecosystems are affected by climate change, this has an impact on perceptions of vulnerability and, thereby, climate action itself (Mazumdaru 2017). The cultural ideas have also, in a way, influenced climate adaptation, wherein the relevance of indigenous knowledge on climate change has been accounted for (Pareek and Trivedi 2011). However, indigenous knowledge has not been adequately integrated into the national action plan and state action plans on climate change (Pallavi 2015).

India's climate diplomacy positions are interspersed with cultural evocations. The distinctions drawn between "environmentalism of the poor" or "livelihood environmentalism" in India and "ecology of affluence" or "full stomach" environmentalism or "middle class lifestyle environmentalism" of the West have influenced India's climate change debates too (Ramesh 2015). In India, most often, an environmental or climate change movement is about establishing a fundamental right to livelihood security and, more importantly, a fundamental right to define the nature and form of development that one aspires for. As Gadgil and Guha (2013) concur, environmentalism began in India at a much early stage of industrialisation, thus rendering it distinct from post-industrial and post-material interpretations of environmentalism. When environmentalism began in the West, environmental quality had already overtaken concerns regarding environmental sustainability, as measures had already been taken to address scarcities through technological and scientific methods. In India, however both problems surfaced at the same time.

Socio-economic and technology ideas

India's perceptions about socio-economic and technological implications of climate change took a turn during this period. They have been shaped partly by India's growing capacities/resources and the interconnected notions of "opportunity", "co-benefits", and so on. In fact, before the Copenhagen Summit, India took up three issues on which it thought there was broad consensus, which were also in India's interest and are generally regarded as "low-hanging fruits" in the negotiations: forestry, CDM, and public–private partnership (PPP) (Ramesh 2009b; Singh et al. 2013). The focus on PPPs was a reflection of India's realisation that demand for exclusive public finance from the developed countries had started to become outdated. Although the NAPCC is more adaptation-centric, its mitigation-focussed parts mostly deal with clean energy in sectors such as power (electricity) and transport (Ramesh 2009b).

These shifts at the domestic level, however, did not necessarily translate into major international commitments of any kind. India was now prepared to include, voluntarily and unilaterally, climate change in the domestic legislative agenda, but not tag it to international commitments, which it maintained would happen only when the industrialised countries took concrete action on this front. Nevertheless, the Indian government identified some areas where it had to "move aggressively in a target oriented manner that would help the country manage GHG emissions without hurting the country's GDP and growth", as asserted by Jairam Ramesh. These areas include mandatory fuel efficiency standards, renewable energy and clean coal expansion, agriculture (to lower methane emissions), carbon sequestration by forests, and mandatory building codes (green-compliant) (Ramesh 2009b). Following the twin-track approach had become the new norm in India's climate diplomacy.

It is, therefore, important to note that on the domestic front, the "co-benefits" approach continued to make an impact in several forums, including the industry and media. This may not have a direct impact on India's climate diplomacy, but it helped shape it through discourse building and normalisation. Navroz Dubash comments that as long as policies had similar end goals, the government was willing to undertake "climate action". For instance, solar mission can be sold as a climate mission in one context and as an energy security measure in another.[4] One of the shifts was a larger focus on building capacity domestically by investing in clean technologies, as technology transfer had progressively become a difficult proposition. Even though the CDM was earlier criticised by India on the basis that offsetting emissions was merely a way of protecting the developed countries' interests and not seriously reducing emissions, the Indian industry, primarily the Confederation of Indian Industry (CII), with the aim of tapping into this potential, supported it and India's participation in it. It is also in consonance with India's position that it would take on mitigation commitments if it received international support. In 2013, the CII, along with WRI India, The Energy and Resources Institute (TERI), launched the India Greenhouse Gas Program (India GHG Program), "a voluntary initiative to standardize measurement and management of GHG emissions in India". Under this initiative, 20 companies have joined hands with environmental and government entities to "promote profitable, sustainable and competitive businesses" (CII 2013).

This increasing interest shown by the Indian industry in influencing climate policy-making has helped build a narrative on climate change as an "opportunity" for Indian companies. This could help them become more efficient, deliver results in terms of Corporate Social and Environmental Responsibility, and become more competitive on the global stage, considering that companies across the world (particularly in the West) have gradually begun to adopt sustainability standards. At the same time, it has contributed to a more "pragmatic" and "result-oriented" approach in India's climate diplomacy. In fact, Ramesh also points out that India could negotiate from a position of strength by accelerating investments in clean technologies domestically and building capacity in various sectors, as technology transfer seldom occurs (Ramesh 2009b).

Political/geopolitical ideas

(Geo)political ideas of "sovereignty", "responsible power", and others have had a long-lasting effect on India's climate diplomacy positions. This is also directly proportional to the international pressure on India to take on binding emissions reduction commitments and its desire to be seen among the "elite" group of major or great powers in international powers. The changes in the climate regime are also noteworthy in this context. The shift from the top-down (targets-and-timetables) approach to a bottom-up (nationally determined contributions) had been preferred by India since the Copenhagen Summit. In this respect, India was supported by not only the BASIC countries but also many developed countries, particularly in the Umbrella Group.[5]

There is, in fact, a direct correlation between India's foreign policy and its climate diplomacy agenda. Some experts see this at times in direct contravention of a science-based approach and its domestic priorities (Raghunandan 2019). Its aspiration of becoming a global player with greater responsibility and stake in governance of global issues hinges on, among other issues, its positions on climate change too. For long, India's uncompromising attitude towards "strategic autonomy" left the country out of materially beneficial partnerships with other countries as well as from international discussion forums on managing the global commons (Mohan 2017). This is a position that it wanted to come out of – which was evident during the Singh administration (and carried forward during the Modi administration too). Climate change became a strategic issue that had got the attention of PM Manmohan Singh and he clearly spelled out the need for India to be a part of the solution, as corroborated by Jairam Ramesh (2009b).

International pressure is yet another factor that needs to be accounted for. To put pressure on the emerging economies, even before the 2009 Copenhagen Summit, a few bilateral meetings were held, including between India and the US. The US and other developed countries (including Denmark, the host) wanted countries such as China and India to either take on emissions reduction targets (potentially binding) or at least announce a cap on their respective emissions. If neither of this was possible, they could declare a peaking year instead. All these demands were unacceptable to India. In effect, the Kyoto Protocol's top-down approach gave way to discussions on the bottom-up model after the Bali Action Summit that would make actions accountable, and not targets (Ramesh 2009b). At the Copenhagen Summit, the Copenhagen Accord was signed, which essentially was one of the first steps to institutionalise this model within the climate regime (Green, Sterner and Wagner 2014). Eventually, it was formally streamlined within the UNFCCC process at the 2010 Cancun Summit, thereby determining the nature of the post-2020 climate regime, which would not be supported by "a treaty-based compliance system", and would be "non-prescriptive" (based on "self-selected nationally determined targets and actions") (Rajamani, Jutta and Doelle 2012).

India's geopolitical ideas of being an influential player that could bring countries, particularly the emerging economies and other developing countries together, are equally important to analyse. At the Copenhagen Summit, India played a major role in stitching up the BASIC group. India's negotiation tactics had changed significantly between the Bali and Copenhagen Summits, knowing that principles such as CBDR-RC and historical responsibility would be watered down, with the result that India would be forced to adopt more commitments. In this context, it would also be crucial to throw light on the institutions, personalities, and other actors that have also contributed to these shifts in climate diplomacy positions. For instance, when Jairam Ramesh was appointed as the Indian Union Minister of State (Independent Charge), Ministry of Environment and Forests (MoEF) in 2009, he became the point person in India's COP delegation. Shyam Saran, who was earlier appointed as the Prime Minister's Special Envoy on climate change had to give way and, with that, the role of diplomats also decreased considerably. However, the Prime Minister's Office's (PMO) grip over climate policy and diplomacy remained untouched, which continues to be the case. In fact, members of the Indian negotiating team, including Saran, were not entirely pleased with the decision to subject India's climate actions to "international consultation and analysis" (recommended by Ramesh under pressure from the US interlocutors). Yet, they supported the announcement of India's national target of reducing the emissions intensity of its GDP by 20–25 percent against 2005 levels by 2020. After the Copenhagen Summit, Saran resigned and the post itself was disbanded, with the MoEF taking charge of the country's climate diplomacy in the negotiations (Varadarajan 2019).

Around the same time, India began to increasingly project itself as a "responsible power" in the climate change negotiations. Ramesh pointed out, "We do not need to be a major emitter to be a major demander of agreement. On climate change, we need to demonstrate vision" (Ramesh 2009b). Having not reneged on any international agreement and claiming the moral high ground on most issues of global importance, the Indian establishment began to see climate change as an "opportunity" to advance its foreign policy agenda – that of gaining more influence in the international system – for which it had to compromise on certain positions. There emerged a feeling that it had to move beyond issues of morality and adopt pragmatic positions, which would fetch India more dividends – both material and ideational. Thus, this was a crucial period in India's climate diplomacy, wherein it had to balance between its framing of climate change as an equity issue (low levels of historical and per capita emissions) and its rise as an emerging economy with rapidly increasing emissions (Saryal 2018).

Moreover, regional isolation (in South Asia) and divisions within the BASIC put further pressure on India. This was another reason why India could not walk alone any longer. India, on the one hand, needed China by its side to influence decision-making, but on the other hand, it remained cautious about being hyphenated with China due to the latter's high-emissions profile. India's isolation got further bolstered by the criticism it faced from its own neighbours, such as the Maldives and Bangladesh, for not committing enough on the climate change front (Jayaram 2019). This is despite the fact that India has always signalled to champion the cause of the developing world's interests. At the same time, China's ability to influence other developing countries and LDCs – particularly among the G-77 – has been substantially boosted by its rapid advancements in renewable energy and its soft power push by pouring money into them (Tan et al. 2013).

Although India's bilateral engagements on climate change were on the rise during this period, whether they had an impact on India's climate diplomacy positions itself is debatable. For instance, with the US, bilateral meetings started shortly after the Copenhagen Summit; and although there was not much progress in the initial years, cooperation began to accelerate since 2013. Climate change was always a backburner issue for a long time and with these bilateral meetings, it became a part of the strategic dialogue. According to Andrew Light, this was possible mainly due to the emphasis being laid on the need for the US to mobilise funds for development-oriented programmes in India, with a focus on solar energy, resilience, and so on.[6] Moreover, India's bilateral cooperation with the other BASIC countries is also worth mentioning, as it could have played a role in building greater understanding between them. China's advancements in renewable energy are known to have acted as a fillip for closer renewable energy cooperation between it and India. A few steps were taken to press forward Sino-Indian relations as specified in the Agreement on Cooperation on Addressing Climate Change (2009) and the Memorandum of Understanding (MoU) on Cooperation on Green Technologies (2010). With China, India's bilateral and multilateral dealings are considered to be guided by shared vulnerabilities and interests (poverty reduction, population growth etc.), and similar national circumstances (developing countries) (Mizo 2016). With Brazil and South Africa too, India has initiated several technological initiatives, both bilaterally and through the trilateral forum of IBSA (signifying South–South cooperation).

The main demand of the IBSA had been the signing of an agreement on the second commitment period of the Kyoto Protocol, besides boosting climate finance by the rich countries (Sharma 2017).

This period saw India becoming more flexible on certain issues related to mitigation commitments and MRV. India's flexibility at the Cancun and Durban Summits was driven by its geopolitical ambitions to be among the permanent members of the UNSC, as some analysts put it. Aligning itself to the US and disengaging with the G-77 on all issues, including climate change, were perceived at that point in time as two important steps that India should take in order to garner the US's support in securing a permanent seat in the UNSC (McDonald and Patrick 2010). Besides, it is evident that India came under international pressure to compromise its positions, even while upholding the most important red lines.[7] It should also be admitted that the evolution in India's positions have been an outcome of its growing understanding of the needs of climate diplomacy, knowledge of other countries' interests, and an ability to impress its interests upon the rest of the international community.

India's negotiating strategy and tactics underwent significant change between the Doha and Paris Summits. It put more efforts into understanding the negotiations process, offering concrete proposals to operationalise principles, and dealing with countries individually in the negotiations, which allowed it to shift from a defensive position to a more proactive one, as commented by Basir Ahmed.[8] Others argue that there continues to be a great amount of uncertainty and a lack of direction in India's climate diplomacy strategy. India continues to use the vulnerability card and project its "emerging power" or "rising power" identity (Jayaram *forthcoming*). While India is increasingly becoming confident, also evidenced by its flamboyant exhibitions and displays of its achievements and goals at the climate summits, its ability to assert itself is also being challenged by the pressure put by the developed countries. At the 2011 Durban Summit, India had to compromise its position on equity and differentiation to some extent. India did not get enough support from the Global South either, which is starkly divided.[9]

To sum up, both ethical and geopolitical ideas dominated India's climate diplomacy discourse, wherein cultural, socio-economic, technological, and environmental/ecological ideas have more or less acted as intervening variables. For example, at the 2013 Warsaw Summit, Jayanthi Natarajan brought up several issues in her statement that became a part of formula narratives of India's climate diplomacy positions in the subsequent years. For instance, she reiterated the sense of consciousness that India had of its "global responsibilities", as much it had that of "equity" and CBDR-RC. Yet, she also remarked, "Poverty eradication stands as our foremost priority. We have huge social and developmental constraints and have to address large unmet energy needs of our vast population". By linking developing countries' post-2020 climate action to developed countries' pre-2020 climate action, India had also stressed the relevance of "high ambition", which should first be shown by the latter for the former to demonstrate the same. By endorsing India's proposal of catalysing and rewarding innovation, and facilitating technology development and transfer, Natarajan also reflected on India's ability and willingness to lead the solution-finding exercise (IEEE 2013). Last but not the least, on a much minor level, the new formula was expanded to include "loss and damage" (explained in detail in Chapter 5), thus reaffirming India's support for the most vulnerable countries and, possibly, its identity as a vulnerable country.

Notes

1 The author's interview with Navroz Dubash, Professor, Centre for Policy Research (CPR), New Delhi, via Zoom on August 10, 2020.
2 The Ad Hoc Working Group on the Durban Platform for Enhanced Action (ADP) is a subsidiary body that was established in December 2011 (Durban Summit). Its mandate is to

> develop a protocol, another legal instrument or an agreed outcome with legal force under the Convention applicable to all Parties, which is to be completed no later than 2015 in order for it to be adopted at the twenty-first session of the Conference of the Parties (COP) and for it to come into effect and be implemented from 2020".

More information is available on the UNFCCC's website: https://unfccc.int/adp-bodies-page.
3 The author's interviews with Navroz Dubash and R. R. Rashmi, Distinguished Fellow and Programme Director, Earth Science and Climate Change, The Energy and Resources Institute; and India's former principal negotiator for climate change negotiations under the UN Framework Convention on Climate Change; and Special Secretary in the Ministry of Environment, Forest & Climate Change in the Government of India via Microsoft Teams on August 17, 2020.
4 The author's interview with Navroz Dubash.
5 The Umbrella Group is a loose coalition of countries, consisting of Australia, Belarus, Canada, Iceland, Israel, Japan, New Zealand, Kazakhstan, Norway, Russia, Ukraine, and the US. It was formed following the adoption of the Kyoto Protocol.
6 The author's interview with Andrew Light, Distinguished Senior Fellow, World Resources Institute, Washington DC, and US Director of the US–India Joint Working Group for Combating Climate Change, via Microsoft Teams on September 4, 2020.

7 The author's interview with Indrajit Bose, senior researcher, Third World Network, via Skype on July 8, 2020.
8 The author's interview with Basir Ahmed, First Secretary, Permanent Mission of India to the UN in Geneva, in Geneva on January 2, 2019.
9 The author's interview with Sandeep Sengupta, global coordinator for climate change at the International Union for Conservation of Nature (IUCN) in Switzerland, in Gland on July 27, 2019. He has previously worked on a wide range of environment and development issues, both within and outside the government in India.

Bibliography

Bikku. 2019. "Religion and Ecological Sustainability among the Bishnois of Western Rajasthan." In *Issues and Perspectives in Anthropology*, edited by Rashmi Sinha, 45–68. New Delhi: Rawat Publications.

Billett, Simon. 2009. "Dividing Climate Change: Global Warming in the Indian." *Climatic Change* 99 (1): 1–16.

Birdsall, Nancy, Subramanian, Arvind, Hammer, Dan and Ummel, Kevin. 2009. "Energy Needs and Efficiency, Not Emissions: Reframing the Climate Change Narrative." *Centre for Global Development*. November. Accessed August 4, 2020. http://citeseerx.ist.psu.edu/viewdoc/download?doi=10.1.1.605.3370&rep=rep1&type=pdf.

Carrington, Damian. 2010. "IPCC Officials Admit Mistake over Melting Himalayan Glaciers." *The Guardian*. January 20. Accessed August 4, 2020. https://www.theguardian.com/environment/2010/jan/20/ipcc-himalayan-glaciers-mistake.

Chakravarty, Shoibal and Ramana, M. V. 2011. "The Hiding behind the Poor Debate: A Synthetic Overview." In *A Handbook on Climate Change and India: Development, Governance, and Climate Change*, edited by Navroz K. Dubash, 218–229. New Delhi: Routledge.

CII. 2013. "India GHG Program Launches with More than 20 Leading Companies." *Confederation of Indian Industry*. July 22. Accessed August 5, 2020. https://www.cii.in/PressreleasesDetail.aspx?enc=XA8vEPiv8W4L0cP3lr ETBvHjSX4FBdEiUfTFxrtvyeY=.

Dasgupta, Chandrashekhar. 2014. "Raising the Heat on Climate Change." *Business Standard*. July 7. Accessed August 4, 2020. https://www.business-standard.com/article/opinion/chandrashekhar-dasgupta-raising-the-heat-on-climate-change-114070701144_1.html.

Dasgupta, Chandrashekhar. 2019. "Present at the Creation: The Making of the Framework Convention on Climate Change." In *India in a Warming World: Integrating Climate Change and Development*, edited by Navroz K. Dubash, 142–156. New Delhi: Oxford University Press.

Dubash, Navroz K. 2009. "Toward a Progressive Indian and Global Climate Politics." *Centre for Policy Research*. September 1. Accessed August 4, 2020. https://www.cprindia.org/research/papers/toward-progressive-indian-and-global-climate-politics.

Dubash, Navroz, Khosla, Radhika, Kelkar, Ulka Kelkar and Lele, Sharachchandra. 2018. "India and Climate Change: Evolving Ideas and Increasing Policy Engagement." *Annual Review of Environment and Resources* 43 (12): 1–30.

Gadgil, Madhav and Guha, Ramachandra. 2013. *Ecology and Equity: The Use and Abuse of Nature in Contemporary India.* New Delhi: Routledge.

Ghosh, Padmaparna. 2008. "India Releases 8-Point Agenda to Combat Climate Change." *Live Mint.* July 1. Accessed August 4, 2020. https://www.livemint.com/Politics/NdI4U5mC8d0tnHGPAqoIhL/India-releases-8point-agenda-to-combat-climate-change.html.

Ghosh, Padmaparna. 2009. "I Want to Position India as a Proactive Player: Jairam Ramesh." *Live Mint.* September 29. Accessed August 4, 2020. https://www.livemint.com/Politics/h97ogi3qEaToXuP9T8YTuI/I-want-to-position-India-as-a-proactive-player-Jairam-Rames.html.

Green, Jessica F., Sterner, Thomas and Wagner, Gernot. 2014. "A Balance of Bottom-up and Top-down in Linking Climate Policies." *Nature Climate Change* 4 (12): 1064–1067.

Hilton, Isabel. 2010. "In India, a Clear Victor on the Climate Action Front." *Yale Environment.* March 1. Accessed August 4, 2020. https://e360.yale.edu/features/in_india_a_clear_victor_on_the_climate_action_front.

IANS. 2011. "India Gets Its Way as Climate Summit in Durban Closes." *Hindustan Times.* December 11. Accessed August 4, 2020. https://www.hindustantimes.com/world/india-gets-its-way-as-climate-summit-in-durban-closes/story-K4NZKLfojcmn0szIz2FlHK.html.

IEEE. 2013. "United Nations Climate Change Conference Warsaw, Poland." *Instituto Español de Estudios Estratégicos.* November. Accessed August 5, 2020. http://www.ieee.es/Galerias/fichero/Varios/Cumbre_CambioClimatico_Varsovia_Nov2013.pdf.

Jayaram, Dhanasree. 2019. "Climate Diplomacy and South Asia: Is There Room for Cooperation?" *India Foundation Journal* 7 (4): 58–67.

Jayaram, Dhanasree. Forthcoming. "India's Climate Diplomacy towards the EU: From Copenhagen to Paris and Beyond." In *EU-India Relations: The Strategic Partnership in the Light of the European Union Global Strategy,* edited by Philipp Gieg, Timo Lowinger, Manuel Pietzko, Anja Zürn, Ummu Salma Bava and Gisela Müller-Brandeck-Bocquet. Cham: Springer. https://indiaclimatedialogue.net/2018/12/03/climate-negotiations-reach-new-crunch-point/.

Lok Sabha Secretariat. 2013. "Climate Change: India's Perspective." *Members' Reference Service* 25.

Mazumdaru, Srinivas. 2017. "Climate Change - India Battles to Balance Economy and Environment." *Deutsche Welle.* November 11. Accessed August 5, 2020. https://www.dw.com/en/climate-change-india-battles-to-balance-economy-and-environment/a-41222773.

McDonald, Kara C. and Patrick, Stewart M. 2010. "UN Security Council Enlargement and U.S. Interests." *Council on Foreign Relations Special*

Report 59. December. Accessed August 9, 2020. https://www.cfr.org/report/un-security-council-enlargement-and-us-interests.

MEA. 2008. "Talk by Special Envoy of Prime Minister, Shri Shyam Saran in Mumbai on Climate Change." *Ministry of External Affairs, Government of India*. April 21. Accessed August 4, 2020. https://mea.gov.in/in-focus-article.htm?18821/Talk+by+Special+Envoy+of+Prime+Minister+Shri+Shyam+Saran+in+Mumbai+on+Climate+Change.

Mehta, Pradeep S. 2011. "History Repeated Itself at Cancun." *Economic Times.* January 17. Accessed August 4, 2020. https://economictimes.indiatimes.com/view-point/history-repeated-itself-at-cancun/articleshow/7300880.cms.

Mizo, Robert. 2016. "India, China and Climate Cooperation". *India Quarterly* 72 (4): 375–394.

MoEF. 2010. "Climate Change and India: A 4X4 Assessment: A Sectoral and Regional Analysis for 2030s." *Indian Network for Climate Change Assessment Report* 2. November. New Delhi: Ministry of Environment and Forests, Government of India.

Mohan, Aniruddh. 2017. "From Rio to Paris: India in Global Climate." *Rising Powers Quarterly* 2 (3): 39–61.

Narain, Sunita, Ghosh, Prodipto, Saxena, N. C., Parikh, Jyoti and Soni, Preeti. 2009. "Climate Change: Perspectives from India." *United Nations Development Programme*. November. Accessed August 5, 2020. https://www.undp.org/content/dam/india/docs/undp_climate_change.pdf.

Pallavi, Aparna. 2015. "How Central Indian Tribes Cope with Climate Change Impacts." *Down to Earth*. August 17. Accessed August 5, 2020. https://www.downtoearth.org.in/news/how-central-indian-tribes-cope-with-climate-change-impacts-43226.

Pandve, Harshal T. 2009. "India's National Action Plan on Climate Change." *Indian Journal of Occupational and Environmental Medicine* 13 (1): 17–19.

Pareek, Aparna and Trivedi, P. C. 2011. "Cultural Values and Indigenous Knowledge of Climate Change and Disaster Prediction in Rajasthan, India." *Indian Journal of Traditional Knowledge* 10 (1): 183–189.

Raghunandan, D. 2019. "India in International Climate Negotiations: Chequered Trajectory." In *India in a Warming World: Integrating Climate Change and Development*, edited by Navroz K. Dubash, 187–204. New Delhi: Oxford University Press.

Rajamani, Lavanya, Brunnée, Jutta and Doelle, Meinhard. 2012. "Introduction: The Role of Compliance in an Evolving Climate Regime." In *Promoting Compliance in an Evolving Climate Regime*, edited by Jutta Brunnée, Meinhard Doelle and Lavanya Rajamani, 1–14. Cambridge: Cambridge University Press.

Ramesh, Jairam. 2009a. "Intervention in the Lok Sabha." Fifteenth Series V. Third Session, 2009/1931. December 3.

Ramesh, Jairam. 2009b. "The Indian Road to Copenhagen." *UChannel*. October 9. Accessed August 4, 2020. https://www.youtube.com/watch?v=LAPxJcyfn0k.

Ramesh, Jairam. 2015. *Green Signals: Ecology, Growth, and Democracy in India*. New Delhi: Oxford University Press.

Ramesh, Jairam. 2017. "Responsibilities of Science, Responsive to Society: A New Dialogue." In *Bridging the Communication Gap in Science and Technology: Lessons from India*, edited by Pallava Bagla and V. V. Binoy, 13–35. Singapore: Springer.

Saran, Samir and Jones, Aled. 2017. *India's Climate Change Identity: Between Reality and Perception*. London: Palgrave Macmillan.

Saran, Shyam. 2008. "Climate Change – From Back Room to Board Room – What Indian Business Needs to Know about India's Approach to Multilateral Negotiations on Climate Change." *Ministry of External Affairs, Government of India*. April 21. Accessed August 4, 2020. https://mea.gov.in/in-focus-article.htm?18821/Talk+by+Special+Envoy+of+Prime+Minister+Shri+Shyam+Saran+in+Mumbai+on+Climate+Change.

Saran, Shyam. 2009. "India's Climate Change Initiatives: Strategies for a Greener Future." *Carnegie Endowment for International Peace*. March 24. Accessed August 5, 2020. https://carnegieendowment.org/files/Saran_Speech%20at%20Carnegie.pdf.

Saryal, Rajnish. 2018. "Climate Change Policy of India: Modifying the Environment." *South Asia Research* 38 (1): 1–19.

Sengupta, Sandeep. 2019. "India's Engagement in Global Climate Negotiations from Rio to Paris." In *India in a Warming World: Integrating Climate Change and Development*, edited by Navroz K. Dubash, 114–141. New Delhi: Oxford University Press.

Sethi, Nitin. 2009. "Jairam for Major Shift at Climate Talks." *The Times of India*. October 19. Accessed August 5, 2020. https://timesofindia.indiatimes.com/india/Jairam-for-major-shift-at-climate-talks/articleshow/5136979.cms.

Sethi, Nitin. 2015. "Journalist Nitin Sethi Answers Readers' Questions on Climate Change and India." *Ecologise*. September 2. Accessed August 4, 2020. https://www.ecologise.in/2015/09/11/nitin-sethi-answers-readers-questions-on-climate-change/.

Sharma, Arundhati. 2017. "India-Brazil-South Africa (IBSA) Trilateral Forum: An Appraisal of Summits." *Indian Council of World Affairs*. December 28. Accessed August 5, 2020. https://www.icwa.in/show_content.php?lang=1&level=3&ls_id=2336&lid=1758.

Singh, Anju, Unnikrishnan, Seema, Naik, Neelima and Duvvuri, Kavita. 2013. "Role of India's Forests in Climate Change Mitigation through the CDM and REDD." *Journal of Environmental Planning and Management* 56 (1): 61–87.

Tan, Xiaomei, Zhao, Yingzhen, Polycarp, Clifford and Bai, Jianwen. 2013. "China's Overseas Investments in the Wind and Solar Industries: Trends and Drivers." *World Resources Institute Working Paper*. April. Accessed August 9, 2020. https://www.wri.org/publication/chinas-overseas-investments-wind-and-solar-industries.

Trading Economics. n.d. "India - Rural Population." *Trading Economics*. Accessed August 4, 2020. https://tradingeconomics.com/india/rural-population-percent-of-total-population-wb-data.html.

Varadarajan, Siddharth. 2019. "Shyam Saran's Exit Suggests Changed Policy Climate." *The Hindu.* February 19. Accessed August 5, 2020. https://www.thehindu.com/opinion/columns/siddharth-varadarajan/Shyam-Saranrsquos-exit-suggests-changed-policy-climate/article16816663.ece.

Varma, Subodh. 2015. "40% of India Still Banks on Monsoon for Agriculture." *The Times of India.* May 1. Accessed August 5, 2020. https://timesofindia.indiatimes.com/india/40-of-India-still-banks-on-monsoon-for-agriculture/articleshow/47115057.cms.

Vihma, Antto. 2011. "India and the Global Climate Governance: Between Principles and Pragmatism." *The Journal of Environment & Development* 20 (1): 69–94.

VOA News. 2010. "India and China: 'Developed Countries Should Lead on Emissions Cuts'." *VOA News.* December 4. Accessed August 4, 2020. https://www.youtube.com/watch?v=g6QqEKYMrsM.

5 The evolution of India's climate diplomacy (2014–2019)

India's climate diplomacy under the Narendra Modi administration with a focus on 2014–2019

Moral/ethical ideas

One of the first moves of the Narendra Modi–led government after coming to power in 2014 was to rename the "Ministry of Environment and Forest" as Ministry of Environment, Forests and Climate Change (MoEFCC). Through this move, the new government had in a way put climate change in the limelight – emphasising its rhetorical commitment to address it more urgently. However, in the climate change negotiations itself, the earlier positions on equity remained non-negotiable. The new administration under Prime Minister (PM) Modi had taken charge by the 2014 Lima Summit. At the summit, Prakash Javadekar, Minister of State (Independent Charge) – Environment, Forest and Climate Change, emphasised the status of "abject poverty" in the world and in India. Opposing the move of developed countries to delink poverty reduction and sustainable development from the post-2020 agreement, he reminded the international community about the indispensability of equity and climate justice in any future climate regime (Aggarwal 2014).

At the Paris Summit, India insisted on the inclusion of climate justice, or as some analysts put it, "climate justice" is merely a "new avatar" of "equity" (Saran and Jones 2017). India's calls for climate justice – a fair share of carbon budget; its attempts to accommodate "differentiation" in the agreed outcome – even in MRV; and its unwillingness to compromise on its coal production goals – 1.5 billion metric tons by 2020 – had created an atmosphere of scepticism in terms of reaching a strong agreement in the beginning (PTI 2015b). During the Paris Summit, India seemed to have compromised on some of its traditionally held red lines. India signed on to an EU proposal for negotiations for a new agreement or legal instrument that did not mention equity, "in a spirit of flexibility and accommodation shown by all" (Down to Earth 2015). At the same time, it also needs to be noted that it was due to the pressure exerted by India that "climate justice" was finally inserted in the Preamble of the agreement (Walt 2015), which was achieved through negotiations held between India and the US.[1]

The environmentalists argued that since the agreement does not mention "historical responsibility", it undermines equity and renders climate action increasingly contingent on "respective capabilities and national circumstances". They also believed that India chose to pay little heed to the absence of strong legal guarantees that developed countries would provide support to developing countries (Narain 2015). The climate law experts, by contrast, insisted that the agreement established a clear relationship between "differentiation", "ambition", and "support" (Dubash 2015). Not only does the agreement reinforce the CBDR-RC principle across all aspects of climate action (mitigation, adaptation and support), but it also ensures that five-year "global stocktakes" take into account "equity" (burden-sharing between countries). Moreover, it contextualises ambitious (future) contributions by developing countries not only in terms of mitigation contributions but also financial contributions by developed countries (Rajamani 2016). Thus, it was also regarded as a diplomatic coup on the behest of India that managed to seal the agreement despite developed countries' opposition. In the words of India's chief negotiator at the Paris Summit, Ashok Lavasa (2019, p. 185):

> I think what worked for India was a firm handling of the key players while keeping some channels open, that is, an accommodating approach on some issues in order to seek flexibility from others on issues that mattered to us, such as: keeping the Montreal Protocol amendments open till the conclusion of Paris Agreement, good rapport with the COP president, support of our traditional allies, tactical moves at the appropriate time, stern messaging and posturing on the penultimate day, liberal interaction with media and NGOs, building a convincing case for development by our logically argued outreach eff orts, and a proactive and positive stance by India throughout that capitalized on the positive and dynamic image of the prime minister.

At the same time, the International Solar Alliance (ISA), launched on the sidelines of the 2015 Paris Summit, by India and France jointly, reinforces the interests of developing countries in pursuing economic development for which "access, availability and affordability of energy" are critical and, in the process, gives impetus to energy transition, riding on the idea of climate justice (Purushothaman 2018).

One of the oft-neglected aspects of the climate change negotiations within the Indian context are the differences among the delegates themselves on issues such as equity and climate justice. Just as differences reportedly simmered between Jairam Ramesh and Shyam Saran at the Copenhagen Summit, such varied opinions are known to have existed on other occasions too.[2] For instance, Pascal Delisle, adviser on climate change at the European External Action Service and co-ordinator of EU Climate Diplomacy Action Plan, comments, "Some [Indian delegates] are more attached to a strong version of CBDR-RC and historical responsibility and/or are extremely sensitive in the way they perceive national sovereignty etc., others are more pragmatic and try to find solutions".[3] In official statements though, these are inalienable aspects. In short, India's commitment to these principles – also tied to the fact that India continues to demand greater "leadership" and climate finance from the developed countries as a part of their pre-2020 obligations – remains steadfast.

Environmental/ecological ideas

India's environmental/ecological vulnerabilities have grown multi-fold in the past few years. Disaster losses in India have skyrocketed in recent decades. Among the top ten global economic loss events in 2019, the southwest monsoon–related floods in India are at the seventh position, with 1,750 deaths and USD 10 billion worth economic loss (AON 2019). However, with time, they are not as starkly represented in the climate diplomacy narratives, except in a few instances such as the Warsaw International Mechanism for Loss and Damage associated with Climate Change Impacts. This mechanism was established at the 2013 Warsaw Summit, at the initiative of the world's most vulnerable developing and least developed countries who had been working towards institutionalising such a mechanism within the UNFCCC as a third pillar (the other two being mitigation and adaptation) for a decade – on the grounds that it is different from adaptation and that they are entitled to compensation (mainly from the developed countries) due to damage(s) caused by climate change (Calliari, Surminski and Mysiak 2018). India, along with the other BASIC countries, has been supportive of this mechanism and has demanded its institutionalisation within the post-2020 climate regime. However, India does not generally use this platform to demand financial support for itself for post-disaster reconstruction, which is a position it has taken since the

2004 Indian Ocean Tsunami. In fact, initially, India was even blamed for being indifferent about this demand (Basu 2013). In addition, this has added one more tenet to South–South cooperation, besides setting the platform for India to launch the Coalition for Disaster Resilient Infrastructure (CDRI) at the 2019 UN Climate Action Summit. This coalition is aimed at "achieving considerable changes in member countries' policy frameworks, future infrastructure investments and high reduction in economic losses from climate-related events and natural disasters across sectors" (CDRI n.d.).

India's extreme climate vulnerabilities also feature in the 2018 Special Report of the Intergovernmental Panel on Climate Change (IPCC) on restricting the temperature rise to 1.5 degrees Celsius. The report was commissioned by the UNFCCC and was released weeks before the 2018 Katowice Summit, wherein the Paris Rulebook was eventually finalised (WRI n.d.). According to the report (prepared by a group of scientists, including Indian ones), agricultural economies such as India would suffer more because of climate change due to droughts, water scarcity, coastal flooding, cyclones, high precipitation events, heatwaves, declining agricultural productivity, rising vector-borne diseases, and extinction of species (IPCC 2018). The IPCC's 1.5 report was acknowledged by India at the national level, wherein the MoEFCC also came up with a proposal to prepare roadmaps for the country to achieve the 1.5 target (instead of the legally accepted 2 degrees Celsius limit at present) (Gupta 2018).

As much as India took cognisance of these vulnerabilities in the domestic sphere, at the international level, most of its focus has been on the fulfilment of the developed countries' pre-2020 targets. While other countries such as the US, Russia, Kuwait, and Saudi Arabia openly opposed the adoption of the IPCC report at the 2018 Katowice Summit, India was accused of being "silent" on the issue. Nevertheless, India's positions are known to be "in sync" with that of the LMDCs as far as the report is concerned – in terms of contextualising the developed countries' pre-2020 commitments incommensurate with the report's findings and call for urgent action (Hemalatha 2018). One of India's delegates, R. R. Rashmi also pointed out, "Any effort to redefine the global goal and fix 1.5 might make it more difficult for those working to maintain under two" (Gupta 2018).

Cultural ideas

Internationally, Indian leaders have repeatedly drawn attention to India's willingness to "take the lead on climate change" on the basis that environmental protection, conservation, and preservation are not new to the Indian ethos (PTI 2018). Prime Minister Modi himself has consistently spoken about the coexistence between humans and nature. In his attempts to shift the narrative from being not just part of the solution but also a potential global climate leader that is capable of setting the agenda on climate-related issues, he has made several proclamations in the run-up to the 2015 Paris Summit. Pointing a finger at the developed countries, he proclaimed that "treating the nature well comes naturally for Indians and they [developed nations] are teaching us" (PTI 2015a). This is, in a way, a form of reinforcement of its civilisational values, which are a part and parcel of its world view. In another case, in a response to President Trump's accusation against India in the context of withdrawing from the Paris Agreement, the Minister of External Affairs Sushma Swaraj asserted, "We signed it because of our belief – a 5,000-year-old belief in environment. If someone says we signed it for money or under pressure, I'll reject it. It is wrong" (Karim 2017).

The use of Sanskrit *shlokas* (poetic forms) by PM Modi and his ministers to pursue certain ideational positions is prominent. At the ISA's founding conference in 2018, Sushma Swaraj ended her address by "bow[ing] before the Sun God and address[ing] him by five different names in Sanskrit" – *Om Suryaya Namaha, Om Adityaya Namaha, Om Dinakaraya Namaha, Om Divakaraya Namaha, Om Bhakasaraya Namaha* ("Oh Sun God! We bow before you who is the Lord of creation, who is the best one to be worshipped, and who illuminates the external and the internal world".) (MEA 2018). Similarly, *Vasudhaiva Kutumbakam* ("the world is one family") is yet another phrase that has often been used by the Modi administration to advance its agenda on multilateralism and global governance (including 'rules-based global order') (FP Staff 2018). Being a part of India's "diplomatic lexicon", it has been used by the previous administrations too in various contexts, although its shortcomings, a lack of strategic vision, and the problems associated with an idealistic world view have been questioned by many critics (Sidhu 2017).

Another major addition to India's narratives on climate action since the Bharatiya Janata Party (BJP)–led National Democratic Government came to power in 2014 is *yoga*. According to Miller (2020), *yoga* has been used by the Modi administration "as a global soft power solution to counter the Global North's climate change privilege on the international stage". Building on Swami Vivekananda's portrayal of *yoga* as a part and parcel of Indian culture and spirituality, the Modi administration has continuously juxtaposed it with the materialistic tendencies of the West. In an address at the United Nations Educational, Scientific and Cultural Organisation (UNESCO) in 2015, PM Modi clearly interlinked India's climate action with *yoga*:

> Yoga awakens a sense of oneness and harmony with self, society and Nature. By changing our lifestyle and creating consciousness, it can help us deal with climate change and create a more balanced world.

In his speech, he reiterates India's commitments as "natural instincts" owing to "culture", "tradition", and "faith" (MEA 2015a). Based on this discourse, India has pointed out that a change in lifestyle (primarily targeted at the West) is necessary for effective climate action. At the 2019 Climate Action Summit too, PM Modi reaffirmed India's commitment to climate action by tying it to India's traditions (Modi 2019):

> What is needed today, is a comprehensive approach which covers everything from education to values, and from lifestyle to developmental philosophy. What we need is a global people's movement to bring about behavioural change.

This spirit of Indian culture and traditions is perhaps a newly reinvigorated discourse in India's foreign policy that has become a cornerstone of India's climate diplomacy positions too.

Socio-economic and technology ideas

India's perceptions of low-carbon development and green transition under the Modi administration did not change much from the previous regime. However, there have been some adjustments in the narrative, combined with innovative framings of climate action itself. In a notable move, the Modi administration placed climate change and the Sustainable Development Goals (SDGs) reached in 2015 side by side. This can be viewed as yet another shift in the framing of climate change in the Indian climate diplomacy discourse. By linking climate change to energy, water, sanitation, healthcare, education, and employment, India had made clear that climate goals cannot and should not be addressed in silos (UNFCCC 2014). Navroz Dubash remarks that Prakash Javadekar used the integration of climate action and SDGs as a strategic move or "marketing strategy" at the international level rather than at the domestic level. Nevertheless, it gives way to technocratic institutions and solutions to address the climate change issue, considering that the SDG framework is essentially technocratic in nature.[4] This can be seen as a redefinition of India's prognostic and motivational framing of climate change at the international level, somewhat reflective of its domestic priorities as well.

The Modi administration continued to restructure the climate debate in order to move away from the traditional perspective of treating climate action and development as competing objectives. For instance, PM Modi talks about moving from "carbon credit" to "green credit" so as to ensure international finance and technology for energy transition through clean/green energy (Jayaram 2015). In his 2015 UNESCO address, PM Modi stated (MEA 2015a):

> Too often, our discussion is reduced to an argument about emission cuts. But, we are more likely to succeed if we offer affordable solutions, not simply impose choices. That is why I have called for global public action to develop clean energy, that is affordable and accessible to all.

This attains more credibility and legitimacy in light of India's announcements to achieve 175 GW of renewable energy by 2022 (out of which solar energy would account for 100 GW) and a series of other commitments such as coal tax, advancement of the national clean energy fund, and so on. With the international community agreeing upon the bottom-up model of climate action (at the UNFCCC), India also declared that it would come up with an "ambitious" Intended Nationally Determined Contributions (INDCs), which had by then become the new catchword in international climate change negotiations (Aggarwal 2015).

Importantly, the tone and tenor of the negotiation tactics showed some signs of change as the Paris Summit neared. One such statement that encapsulates this change is that of the former Indian ambassador to the US, Arun K. Singh. According to him, India is not a contributor to either historical or current emissions (on account of its low per capita emissions). However, in view of the consequences of climate change which are being felt by the country and become of the deteriorating global situation, the Indian government is determined to develop its economy, "to the maximum extent possible", based on renewable energy. This would happen as India develops at the rates of 8–10 percent (necessary for poverty reduction in the country). The stress on energy security is further emphasised by him through the instance of "more than 300 million Indians" without "access to commercial energy" or electricity (VOA News 2016). From the above statement, it is clear that the Indian position on climate action based on historical responsibility softened marginally and that its position on commitments would be defined by action on the front of expansion of renewable energy, and not so much by the unspecified international obligation of reducing GHG emissions. It is merely a difference in perception, as the former anyway leads to the latter.

One of the biggest selling points of the Modi administration is PM Modi's personal interest in the issue of climate change. Considering that PM Modi himself has authored two books on climate change and highlighted the need for addressing it urgently even while serving as the Chief Minister of the state of Gujarat, hopes have been high that India would be more proactive when it comes to climate action under his leadership (Modi 2010). His second book *Convenient Action: Continuity for Change* was released during the Paris Summit alongside the ISA's launch. The book outlines his proposal of "building on a model of positive partnership between people, businesses, scientific community, government and NGOs" with a focus on "delivering clean energy and prosperity through site-dependent initiatives and scaling up efforts to make rapid transformation" (Modi 2015). However, critiques have pointed out several loopholes in his perceptions about climate change and action, including mixing up of concepts (such as "local pollution problems" being "mixed up with carbon emissions") and selling of controversial solutions such as dams (Sethi 2014).

Irrespective of these ambivalent positions, domestically, many goals were declared before the Paris Summit, targeted mainly at clean energy, even before the Indian government submitted INDCs[5] to the UNFCCC. For instance, former Minister of Finance Arun Jaitley stated in his budget speech in 2015 that India planned to "quadruple its renewable power capacity to 175 gigawatt (GW) by 2022", with 100 GW alone as being proposed to be sourced from solar, 60 GW from wind, 10 GW from biomass, and 5 GW from hydroelectricity (Nagarajan 2015). At the same time, before India submitted its INDC, it was speculated that India could opt to submit two INDCs to the UNFCCC – one that could be achieved with domestic resources and the other with the help of financial and technological resources from the developed countries (Gupta 2015). India has had differences with the developed countries over the preoccupation of the climate regime with mitigation without giving due importance to adaptation, climate finance, and loss-and-damage mechanisms.

Domestically, many more initiatives have been announced, which reinforce a continued focus on low-carbon development in some sectors. In fact, it is not just the renewable energy sector that India is concentrating on, albeit it is the most prominent one. The Modi administration has rolled out a series of policies that are directed at reducing the energy intensity of India's cities, transport, and infrastructure. For example, Green Energy Corridors and the National Smart Grid Mission are aimed at increasing the country's "energy capacities from wind and waste conversion" (ET Contributors 2017). Programmes related to health, soil health management, coastal management, watershed management, irrigation systems, organic farming, and climate adaptation among others are also proposed or are in the pipeline. These are in addition to the eight missions of the NAPCC (Jayaram 2017).

The "Make in India" programme, launched by Prime Minister Modi in 2014, is an initiative aimed at transforming India into a global manufacturing hub through investments and innovation (Jayaram *forthcoming*). It may be seen as anti-climate agenda due to its emphasis on the manufacturing sector. However, the government also seems to be keen on mainstreaming a few green elements in the initiative through efforts such as inviting investments in clean energy or incorporating the principle of "zero defect, zero effect" etc. (Mohan 2017; Venugopal 2016). Another oft-discussed flagship programme of the Indian government is the Smart Cities Missions that aims to transform Indian cities into spaces that are more clean, sustainable, inclusive, and people-friendly. These actions are linked to its climate action goals as well as SDGs. These initiatives are yet again not devoid of shortcomings that have been covered by many analysts in the media and otherwise, but rhetorically, they carry weight as India attempts to strengthen its place in the international climate order by showing what is being done on the domestic front to back its proclamations at the international level.

Programmes such as these are certainly aiding India's cooperation with other countries in the field of clean technology. For example, India is looking to leverage Sweden's leadership in "urban mobility solutions, smart parking systems, air filtration, waste management solutions and real time information systems". To this effect, a Memorandum of Understanding (MoU) between the two countries on Sustainable Urban Development was signed in 2015. Some of the proposed innovative interventions include CALE parking solutions, Scania's ethanol-run "Green Bus", Volvo's hybrid buses, ABB Solution Network Manager SCADA's energy efficiency solutions, and Fortum's clean energy solutions (Jayaram *forthoming*; Make in India 2018).

The ISA is yet another international initiative through which India wishes to enhance cooperation with other countries on solar energy and usher in material advancement in this sector domestically. Apart from climate action goals, this institution is also aimed at facilitating India's attempts to fulfil SDGs, especially SDG-7 on energy (MNRE 2015). This is congruent with India's new formula of tying climate action with SDGs (Bhattacharya, Niranjan and Purushothaman 2018; Jayaram *forthcoming*). Both India and France have expressed their keenness to set the ball rolling to achieve the ISA's 1000 GW target by 2030 globally, despite the fact that many have expressed scepticism regarding the coalition's ability to mobilise the required funds to the tune of US$1 trillion (Kumar 2018). Even though the goals are ambitious and the solar sector globally is fraught with challenges, the overall mood is optimistic. This alliance can be viewed as an attempt to transform the post-2020 climate order and align it with the changing geopolitical realities by providing solutions that build on individual countries' strengths. The ISA[6] envisages to mobilise countries with plentiful sunshine (but lack resources) and those that have a comparatively larger pool of technological, financial, and technical resources (Purushothaman 2018).

One of the biggest bones of contention in the post-2020 climate regime is the future of the market-based mechanisms, as already described in the previous chapters too. One of India's biggest worries is that it could lose all the carbon credits it has earned over the years under the Kyoto regime. However, most countries are not in favour of continuation of the credits, except a few fast-growing (emerging economies) and developed countries. The existing trust deficit between the developed countries and emerging economies has further exacerbated the dispute, and no agreement could be reached until 2019, despite urgency (Khadka 2019). At the 2019 BASIC meeting, it was unanimously decided by the four countries that the international community should ensure a smooth transition of the CDM, from which they have benefited immensely (especially China and India). The joint statement called for creating an environment for investments and businesses, involving both public and private players – thereby embracing the "development" paradigm, with "environmental integrity", presumably meant to ensure that it is sustainable too (PIB 2019b).

Political/geopolitical ideas

The change in the name of the environment ministry in a way indicates the growing consciousness about India's responsibility towards tackling an issue of global governance (Saryal 2018). It is here that Modi's *Panchamrit* (considered a replacement of *Panchsheel* in India's foreign policy agenda) also needs to be highlighted, which includes the five themes of *samman* ("dignity and honour"), *samvad* ("greater engagement and dialogue"), *samriddhi* ("shared responsibility"), *suraksha* ("regional and global security"), and *sanskriti evam sabhyata* (cultural and civilisational linkages") (Mohan, Archis 2015). Not only is *Panchamrit* (five nectars) steeped in the tradition and practice of multilateralism, but it also exudes cultural and civilisational values and "national pride" (Chaturvedi 2016; Hombal 2015). India's priorities in terms of global governance are enshrined through these principles and they are aimed at achieving "transparency", "credibility", "accountability", and "effectiveness". They also reinforce India's commitment towards the "rule of law" and "a fair and equitable international system" (MEA 2020). These principles have, therefore, become a cornerstone of India's climate diplomacy too.

Where bilateral relations started to assume even more significance was when the US began to take active interest in leading the negotiations at the Paris Summit, months before it. It reached out to both China and India for a consensus deal. In the meetings that former US President Obama had with Chinese President Xi and Indian PM Modi, the climate diplomacy positions of India were amplified. China's joint communiqué with the US, specifying its year of peaking emissions, had political reverberations that were felt in India too. Shyam Saran pointed out that if China had agreed to peak its emissions in 2030, then India could agree to do the same 15–20 years after 2030, taking into account India's aggregate and per capita emissions that are far lower than that of China (Mohan and Bagchi 2014).

India's "flexibility" on the global stage was further enunciated by its reciprocal pact with the US. During Obama's visit to India in 2015, months before the Paris Summit, the two countries made headway on the issue of hydrofluorocarbons (HFCs). This was considered a "breakthrough" (Neuhauser 2015), since India had for long opposed moving the HFCs from the climate regime to the ozone regime (Montreal Protocol on Substances that Deplete the Ozone Layer). Since HFCs are not Ozone Depleting Substances (ODS) and are being used as an alternative to the ODS, the Indian government had reservations about amending the Montreal Protocol to address HFCs, which are a potent GHG whose Global Warming Potential ranges from 12 to 14,800. Although India did not change this stance in the Montreal Protocol meeting that succeeded the Modi–Obama meeting, it finally signed the Kigali amendment to the Montreal Protocol in 2016 that is aimed more at addressing climate change.[7]

On the question of "international pressure" however, India continued to uphold the principles of sovereignty and strategic autonomy.

Although a "concrete" agreement (as opposed to China) was not reached in terms of GHG emissions reduction commitments, a few crucial steps were taken in the direction of securing a stronger agreement at the Paris Climate Summit later in 2015. PM Modi's statement during the joint press conference held by the two leaders made it clear that India would not only continue to place equal emphasis on adaptation and climate finance (besides mitigation) in the negotiations but also avoid any commitments aimed at peaking emissions. Although he underscored the sense of urgency and severity associated with climate change, he also asserted that India, being a sovereign country, cannot be pressurised by any external force to take on commitments (alluding to China's agreement with the US). Nevertheless, he refrained from emphasising too much the traditionally held positions such as the CBDR-RC and instead expressed willingness to work towards reaching a robust global agreement at the Paris Summit – which in itself can be seen as a shift in India's approach towards global climate policy (Jayaram 2015).

The two countries successfully concluded an MoU on Energy Security, Clean Energy, and Climate Change. Besides, they also launched cooperative mechanisms to improve air quality, climate resilience, and capacity-building (such as through the Fulbright–Kalam Climate Fellowship) (Mohan, Vishwa 2015). Clearly, the preoccupation with the treaty language and what goes into the post-2020 climate agreement gave way to finding avenues of cooperation at the practical level. According to Andrew Light, the US began to push for cooperation through multiple agencies – the White House, US State Department, the US Embassy (in New Delhi), the Bureau of South and Central Asian Affairs, and others. This added teeth to Indo–US climate cooperation at various levels. Interestingly, the expectation within the US was that a strong bilateral relationship with India on climate change could translate into similar such arrangements with Brazil and South Africa too, thus helping advance global climate goals at the multilateral level too.[8]

The personality factor and PM Modi's "massive outreach" (in particular, in the run-up to the Paris Summit) to the rest of the world cannot be left out of an analysis of India's climate diplomacy (Lavasa 2019). As much as PM Modi has focussed on promoting India's multilateral agenda, he has also used "bilateralism" to advance India's interests. He is known for his frequent foreign trips that have resulted in many bilateral agreements and statements since 2014. On the one hand, these visits are expected to raise "India's global stature", and on the other, they have reportedly "brought in a flood of investment"

(Bloomberg 2019). The majority of these visits have featured climate change and clean energy as important areas of discussion and cooperation. His visits to Germany, France, and Canada in 2015 before the Paris Summit also helped set the tone for India's willingness to be an agenda-setter on the global stage (PTI 2015a).

India's bilateral climate diplomacy with major players such as the European Union (EU), and especially some member states within the EU in terms of hands-on cooperation has been on the rise. An example is the case of Germany, with which India has launched various initiatives such as the Indo-German Working Group on Climate Change under the Indo-German Environment Forum (MEA 2015c). Yet, the bilateral relations between India and the EU are considered complicated. The EU's long-standing, principled approach based on precaution clashes with India's insistence on equity. However, the EU's insistence on reaching an ambitious agreement in Paris and India's need for not being seen as a "spoiler" led to ideational convergence and cooperative behaviour between the two parties at the Paris Summit too (Belis et al. 2018). More importantly, as pointed out by Pascal Delisle, "We have moved from a very diplomatic type of interaction needed to get to a universal agreement/Paris Agreement to a more concrete and more balanced position where a big part of our interaction is about solutions, International Solar Alliance, peer to peer cooperation and so on". He also acknowledged that despite growing convergences between India and the EU, it cannot be compared to the EU–China climate diplomacy, which is developing much faster with greater cooperation in carbon and energy markets.[9] Interestingly, although India has cooperated with China through forums such as the BASIC as far as the climate change negotiations are concerned, it has preferred bilateral hands-on cooperation with the West. It believes that there is better scope for building capacities domestically by leveraging the West's leadership in cutting-edge technology.[10]

While India's collaborations with the rich countries have been on the rise, it has also begun to provide financial support to other developing and least developed countries. India has signed multiple bilateral and multilateral deals with countries to enhance investments in clean energy, as well as assist other countries in tackling the effects of climate change. For instance, India has facilitated the launch of a project with the United Nations (UN) to develop climate early warning systems for building resilience in seven Pacific island countries – Tonga, Kiribati, the Marshall Islands, Solomon Islands, the Cook Islands, Nauru, and the Federated States of Micronesia (RNZ 2017). Similarly, India's cooperation with Africa in the area of climate action and sustainability has increased. In 2019, the India and Africa Partnership for Sustainability was hosted on the sidelines of the United Nations Environment Assembly (UNEA) (Kaul 2019). In fact, India has also provided funding to researchers from several countries of Africa to conduct research in various areas, including renewable energy, food security, and so on (MEA 2015d). In another instance, the India–China Joint Statement on Climate Change (2015) is focussed upon strengthening "practical bilateral cooperation, including in areas of clean energy technologies, energy conservation, energy efficiency, renewable energy, sustainable transportation including electric vehicles, low-carbon urbanization and adaptation". (MEA 2015b).

India's climate diplomacy has also been partly driven by the type of isolation that India has faced on several occasions in the climate change negotiations. Not only have countries within the BASIC isolated India when it came to decisions on MRV, second commitment period of the Kyoto Protocol, and equity, but other developing and least developed countries have also joined hands with parties such as the EU to extract concessions from India (in effect, isolating it), thereby diluting some of India's red lines.[11] India's isolation was further increased by the joint announcement of Barack Obama and Xi Jinping in 2014 – a preparatory initiative ahead of the Paris Summit – emboldening the G-2 (China and the US on global governance) platform. Yet as some say, on issues such as trade and climate change, India and China have been on the same side of the table. Even though there have been speculations that the gap between the two have widened due to China's growing emissions (and consequently more pressure on it to act) and changing positions as the Paris Summit neared, many confirm that the cooperation between the two Asian giants is still intact.[12]

At the Paris Summit, India was the centre of attention, as both the industrialised and least developed countries saw India as a "challenge", as described by former US secretary of state John Kerry (Ananthakrishnan 2015). Interestingly, as the Paris Summit drew closer, India's tag of being a "spoiler" had slowly given way to that of a "bridging" nation – bridging "the many nations across the world" as well as bridging "development with climate action" in the media narratives (Jaiswal 2015). Nevertheless, India's attempts to make all the right noises owing to various international and domestic dynamics were stark. India did not want to be seen as a "bad boy", especially in light of China's agreement with the US and its willingness to provide green finance to developing countries, which India was not in a position to announce (Lavasa 2019).

Another milestone that has boosted India's image in the global climate order is the launch of the ISA at the Paris Summit, which became a legal entity on December 6, 2017. It is considered to be a major diplomatic victory for India (Mohapatra 2019). The ISA is not just a sign of India's "flexibility" in terms of its desire to be a part of the solution and seek partnerships, but it is also an indication of India's attempts to catapult itself to the position of a global climate "leader". The ISA is also seen as a natural evolution of PM Modi's role as the Chief Minister of the state of Gujarat, where he worked on advancing solar energy, water security, and so on, according to Andrew Light.[13] The fact that the ISA is the first "treaty-based international government organisation to be based in India" is noteworthy (Bhaskar 2017). Through the ISA and unswerving support for the Paris Agreement (through several statements), India began to elevate its position in the international climate order. Paul Watkinson comments that the ISA was "strongly moved forward by Prime Minister Modi", based on "concrete outputs", and "identifying the real need in the tropics and sub-tropics for access to technology".[14] One could even argue that the ISA specifically gave India a place on the global high table, at least in the realm of international climate policy. The ISA, as some analysts put it, "encapsulates the spirit of the Paris Agreement: what every country can do, and how we can do better together" (Goswami 2018; Jayaram *forthcoming*).

Through the ISA too, India has sought to strengthen ties with other developing countries in Africa, Asia, Latin America, and the Pacific. The ISA has emerged as an integral part of India's foreign policy agenda. Delisle, for instance, remarks that "the launch of the ISA in Paris has allowed India to increase its profile as a possible global leader, or a part of the solution moving from a position of having the right to pollute like others did before, to a more lucid position of what India's interests are in a global economy".[15] Ideationally, PM Modi has advocated the ISA through slogans and messages such as "One World, One Sun and One Grid" (reminiscent of *Vasudhaiva Kutumbakam* (MEA 2018) – The entire world is our family) and "alternative OPEC" (guided by material factors) (Mohapatra 2019). Hence, for India, it is a twin-pronged strategy – first, to rejuvenate its solar sector that has become somewhat stagnant due to various reasons, including import duty on solar modules and flat power demand, and which needs fresh investments (Tripathi 2018); and second, to increase its sphere of influence through soft power. For instance, India announced $1 billion assistance for 23 solar power projects across 13 countries in Africa by offering solar panels manufactured by Indian companies at a lower cost than Chinese ones (Saurabh 2018).

The most recent initiative of launching the Coalition for Disaster Resilient Infrastructure at the 2019 UN Climate Action Summit can be looked upon as yet another way of signifying India's desire to ignite a global movement and boost its image of a global leader (PIB 2019a). India's strategy of targeting specific issues to build a coalition – first, energy and then, disasters – can be understood through two approaches. First, it is now interested in concrete solutions that may not entirely depend on the status of the international climate change negotiations. This could possibly help it overcome the UNFCCC process, in which its position has been precarious. Second, these issues are also tied to India's own interests, thus bringing forth the importance of two principles of climate diplomacy – knowing one's own interests and the capacity to carry "ideas" through to fruition (Mabey, Gallagher and Born 2013).

During 2015–2019, India is known to have hardened its stance on many issues, which could be attributed to many reasons – including the lack of financial commitments from the developed countries, President Trump's decision on US's withdrawal from the Paris Agreement, China's rise as a global powerhouse, as well as divisions among the Indian delegations (in the negotiations).[16] As much as India kept harping about the pre-2020 commitments (that remained unfulfilled), including the fact that the Doha Amendment[17] had not been enforced (by 2019), at the 2019 Madrid Summit, Prakash Javadekar clarified in his statement, "India has reduced emission intensity of GDP (gross domestic product) by 21 percent and is on track to achieve the goal of 35 percent emission reduction as promised in Paris". He reiterated India's vision of "walking the talk" in this respect as well. It is one of the few countries that are in a position to achieve their NDC targets; and in fact, India is leading the list (PTI 2019).

Therefore, ideationally, India is now balancing between its vision of "strategic autonomy" and allegiance to a "rules-based global order", driven by mutual interests and promotion of "security, stability and sustainability", as reinforced by PM Modi during his bilateral visit to Germany in 2017 (MEA 2017). At the same time, some analysts believe that through partnerships and collaborations, India seeks to replace the Western model of zero-sum to a win-win model, which is a necessity in global climate governance based on climate ethics and climate justice.[18] The country's climate diplomacy is, therefore, understood by many analysts (as explained in this chapter) to have evolved from a rather confrontational stance in the Copenhagen period to that of a more cooperative player in the Paris period. For India, climate change represents a cornerstone of the Modi government's reinvigorated foreign policy that strives to project India as a power that has already "risen" and whose influence cannot be restricted to the subcontinent.

Notes

1 The author's interview with Andrew Light, Distinguished Senior Fellow, World Resources Institute, Washington DC, and former US Director of the US–India Joint Working Group for Combating Climate Change, via Microsoft Teams on September 4, 2020.
2 The author's interview with Navroz Dubash, Professor, Centre for Policy Research (CPR), New Delhi, via Zoom on August 10, 2020.
3 The author's interview with Pascal Delisle, Adviser on Climate Change at the European External Action Service and Coordinator of EU Climate Diplomacy Action Plan, in Brussels on February 15, 2019.
4 The author's interview with Navroz Dubash.
5 The Intended Nationally Determined Contribution (INDC) constitutes the publicly outlined post-2020 climate actions that countries intended to take under the Paris Agreement. The INDC (communication) becomes NDC or Nationally Determined Contribution, once a country submits its respective instrument of ratification, accession, or approval to join the Paris Agreement. More information is available on the World Resources Institute's website: http://www.wri.org/indc-definition.
6 More information about the International Solar Alliance is available on the website: https://isolaralliance.org/.
7 Under this amendment, the rich countries or non–Article 5 countries with the exception of a few are expected to start to reduce production by 10 percent in 2019. Among the Article 5 countries or developing countries, the majority of countries are required to phase down the production of HFCs in 2024. However, India is among those countries that have a later freeze date and are expected to phase down the production only in 2028. More information is available on the UNEP's website: http://www.unep.fr/ozonaction/information/mmcfiles/7809-e-Factsheet_Kigali_Amendment_to_MP.pdf.
8 The author's interview with Andrew Light.
9 The author's interview with Pascal Delisle, Adviser on Climate Change at the European External Action Service and Coordinator of EU Climate Diplomacy Action Plan, in Brussels on 15 February 2019.
10 The author's interview with Shyam Saran, former Special Envoy of the Indian Prime Minister on Climate Change (SEPM), in New Delhi on 15 June 2011.
11 The author's interview with Indrajit Bose.
12 The author's interviews with Basir Ahmed and Indrajit Bose.
13 The author's interview with Andrew Light.
14 The author's interview with Paul Watkinson, one of the EU's lead negotiators at the conferences of Copenhagen, Cancun, Durban, Doha, and Warsaw; and chief negotiator and head of the climate negotiations team for the French ministry of Ecology, Sustainable Development and Energy, via Skype on January 17, 2019.

15 The author's interview with Pascal Delisle.
16 The author's interviews with Navroz Dubash; Indrajit Bose, senior researcher, Third World Network, via Skype on July 8, 2020; and Andrew Light.
17 More information on the Doha Amendment (with reference to the second commitment period of the Kyoto Protocol is available on the UNFCCC's website: https://unfccc.int/process/the-kyoto-protocol/the-doha-amendment).
18 Based on views expressed by M. D. Nalapat, Editorial Director of Itv network (India) and The Sunday Guardian and UNESCO (United Nations Educational, Scientific and Cultural Organization) Peace Chair, in a panel discussion on "India's Leadership in Times of Environmental Change" at Manipal Academy of Higher Education (MAHE), on April 4, 2016.

Bibliography

Aggarwal, Mayank. 2014. "India against Keeping Poverty Reduction Priority Out of Climate Pact." *Live Mint.* December 11. Accessed August 6, 2020. https://www.livemint.com/Politics/53e6uWjUTWHEqUztPkJ0fL/India-against-keeping-poverty-reduction-priority-out-of-clim.html.

Aggarwal, Mayank. 2015. "Replace Carbon Credit with Green Credit: Narendra Modi." *Live Mint.* December 2. Accessed August 9, 2020. https://www.livemint.com/Politics/V7FI8fMqpRVywH5RS2pl6M/Replace-carbon-credit-with-green-credit-Narendra-Modi.html.

Ananthakrishnan, G. 2015. "Kerry Finds India 'Positive' at Paris Climate Conference." *The Hindu.* December 9. Accessed August 9, 2020. https://www.thehindu.com/sci-tech/energy-and-environment/cop21-kerry-finds-india-positive-at-paris-climate-conference/article7962418.ece.

AON. 2019. "Weather, Climate & Catastrophe Insight." *AON Annual Report.* Accessed August 6, 2020. https://www.aon.com/global-weather-catastrophe-natural-disasters-costs-climate-change-2019-annual-report/index.html?utm_source=regionemeauk&utm_campaign=natcat20.

Basu, Jayanta. 2013. "India Sides with Rich Nations on Climate Damage Compensation." *Earth Journalism Network.* November 18. Accessed August 9, 2020. https://earthjournalism.net/stories/india-sides-with-rich-nations-on-climate-damage-compensation.

Belis, David, Schunz, Simon, Wang, Tao and Jayaram, Dhanasree. 2018. "Climate Diplomacy and the Rise of 'Multiple Bilateralism' between China, India and the EU." *Carbon & Climate Law Review* 12 (2): 85–97.

Bhaskar, Utpal. 2017. "International Solar Alliance Becomes India's Calling Card on Climate Change." *Live Mint.* December 6. Accessed August 9, 2020. https://www.livemint.com/Industry/97DgLussreA6CXvmH7LVQM/International-Solar-Alliance-becomes-Indias-calling-card-on.html.

Bhattacharya, Shreya, Niranjan, Ekta and Purushothaman, Chithra. 2018. "India and the International Solar Alliance." *IDSA Backgrounder.* March 16. Accessed February 20, 2020. https://idsa.in/system/files/backgrounder/b-india-international-solar-alliance_shreya.ekta_.chithra_160318.pdf.

Bloomberg. 2019. "92 Trips, 57 Countries: What PM Modi Achieved on His Foreign Visits." *The Times of India.* April 30. Accessed August 7, 2020. https://timesofindia.indiatimes.com/india/92-trips-57-countries-what-pm-modi-achieved-on-his-foreign-visits/articleshow/69106996.cms.

Calliari, Elisa, Surminski, Swenja and Mysiak, Jaroslav. 2018. "The Politics of (and Behind) the UNFCCC's Loss and Damage Mechanism." In *Loss and Damage from Climate Change: Concepts, Methods and Policy Options,* edited by Reinhard Mechler, Laurens M. Bouwer, Thomas Schinko, Swenja Surminski and JoAnne Linnerooth-Bayer, 155–178. London: Springer.

CDRI. n.d. "About CDRI." *Coalition for Disaster Resilient Infrastructure.* Accessed August 8, 2020. https://cdri.world/coalition-for-disaster-resilient-infrastructure.php.

Chaturvedy, Rajeev Ranjan. 2016. "India's Foreign Policy and the 'Modi Doctrine'." *Institute of South Asian Studies Special Report* 38. November 2. https://www.isas.nus.edu.sg/wp-content/uploads/media/isas_papers/ISAS%20Special%20Report%20No.%2038.pdf.

Down to Earth. 2015. "Time Out." *Down to Earth.* June 7. Accessed August 5, 2020. https://www.downtoearth.org.in/coverage/timeout-34650.

Dubash, Navroz K. 2015. "A Climate More Congenial to India." *The Hindu.* December 16. Accessed August 5, 2020. http://www.thehindu.com/opinion/op-ed/cop-21-and-paris-agreement-a-climate-more-congenial-to-india/article7992802.ece?homepage=true.

ET Contributors. 2017. "A Paradigm Shift on Climate Change under PM Modi, Says Harsh Vardhan." *The Economic Times.* June 5. Accessed August 8, 2020. https://economictimes.indiatimes.com/news/politics-and-nation/a-paradigm-shift-on-climate-change-under-pm-modi-says-harsh-vardhan/articleshow/58991578.cms.

FP Staff. 2018. "'India Offers Everything You Seek': Full Text of Narendra Modi's WEF 2018 Opening Plenary Address." *First Post.* January 24. Accessed August 9, 2020. https://www.firstpost.com/world/world-economic-forum-davos-narendra-modi-address-read-full-text-india-is-an-investment-in-the-future-full-text-of-narendra-modis-speech-at-plenary-session-of-wef-4316305.html.

Goswami, Urmi A. 2018. "International Solar Alliance: India's Place under the Sun." *The Economic Times.* March 13. Accessed August 9, 2020. https://economictimes.indiatimes.com/news/science/international-solar-alliance-indias-place-under-the-sun/articleshow/63290304.cms.

Gupta, Joydeep. 2015. "India Offers Two Options for UN Climate Deal." *India Climate Dialogue.* February 3. Accessed August 9, 2020. https://indiaclimatedialogue.net/2015/02/03/india-offers-two-options-un-climate-deal/.

Gupta, Joydeep. 2018. "India Seeking Ways to Limit Climate Change after IPCC Report." *The Third Pole.* October 29. Accessed August 7, 2020. https://www.thethirdpole.net/2018/10/29/india-ipcc/.

Hemalatha, Karthikeyan. 2018. "Climate Summit Fails to Adopt Key Report as US Pitches Fossil Fuels, India and Australia Keep Mum." *Earth Journalism.* December 13. Accessed August 8, 2020. https://earthjournalism.

net/stories/climate-summit-fails-to-adopt-key-report-as-us-pitches-fossil-fuels-india-and-australia.

Hombal, N. B. 2015. "BJP Describes PM's Foreign Policy as 'Panchamrit'." *Deccan Chronicle.* April 4. Accessed August 9, 2020. https://www.deccanchronicle.com/150404/nation-current-affairs/article/bjp-describes-pm%E2%80%99s-foreign-policy-%E2%80%98panchamrit%E2%80%99.

IPCC. 2018. "Summary for Policymakers." In *Global Warming of 1.5°C. An IPCC Special Report on the Impacts of Global Warming of 1.5°C above Pre-industrial Levels and Related Global Greenhouse Gas Emission Pathways, in the Context of Strengthening the Global Response to the Threat of Climate Change, Sustainable Development, and Efforts to Eradicate Poverty,* edited by V. Masson-Delmotte et al. Accessed August 5, 2020. https://www.ipcc.ch/sr15/chapter/spm/.

Jaiswal, Anjali and Connolly, Meredith. 2015. "Paris Climate Agreement Explained: India Focus." *Natural Resources Defense Council.* December 12. Accessed August 9, 2020. https://www.nrdc.org/experts/anjali-jaiswal/paris-climate-agreement-explained-india-focus.

Jayaram, Dhanasree. 2015. "A Shift in the Agenda for China and India: Geopolitical Implications for Future Climate Governance." *Carbon & Climate Law Review* 9 (3): 219–230.

Jayaram, Dhanasree. 2017. "India: The Next Global Climate Leader?" *Science, Technology and Security Forum.* June 19. Accessed August 9, 2020. http://stsfor.org/content/india-next-global-climate-leader.

Jayaram, Dhanasree. Forthcoming. "India's Climate Diplomacy towards the EU: From Copenhagen to Paris and Beyond." In *EU-India Relations: The Strategic Partnership in the Light of the European Union Global Strategy,* edited by Philipp Gieg, Timo Lowinger, Manuel Pietzko, Anja Zürn, Ummu Salma Bava and Gisela Müller-Brandeck-Bocquet. Cham: Springer.

Karim, Meeran. 2017. "Indian Prime Minister to Visit Trump, and a Few Topics Will Be Tense." *Slate.* June 12. Accessed August 5, 2020. https://slate.com/news-and-politics/2017/06/indian-prime-minister-narendra-modi-to-visit-president-trump-here-s-what-they-ll-talk-about.html.

Kaul, Aastha. 2019. "The India-Africa Partnership for Sustainability." *Observer Research Foundation Special Report.* April 7. Accessed August 5, 2020. https://www.orfonline.org/research/special-report-the-india-and-africa-partnership-for-sustainability-49334/.

Khadka, Navin Singh. 2019. "Paris Agreement: Will India Lose Millions of Carbon Credits?" *BBC News.* December 26. Accessed August 9, 2020. https://www.bbc.com/news/world-asia-india-50774901.

Kumar, Sanjay. 2018. "India and France Pledge Billions of Dollars on Solar Energy." *Nature.* March 14. Accessed August 8, 2020. https://www.nature.com/articles/d41586-018-03126-3.

Lavasa, Ashok. 2019. "Reaching Agreement in Paris: A Negotiator's Perspective." In *India in a Warming World: Integrating Climate Change and*

Development, edited by Navroz K. Dubash, 169–186. New Delhi: Oxford University Press.

Mabey, Nick, Gallagher, Liz and Born, Camilla. 2013. *Understanding Climate Diplomacy: Building Diplomatic Capacity and Systems to Avoid Dangerous Climate Change*. London: Third Generation Environmentalism.

Make in India. 2018. "Swedish Solutions for Indian Smart Cities." *Make in India*. Accessed August 9, 2020. https://www.makeinindia.com/article/-/v/swedish-solutions-for-indian-smart-cities.

MEA. 2015a. "Prime Minister's Address to UNESCO." *Ministry of External Affairs, Government of India*. April 10. Accessed August 8, 2020. https://mea.gov.in/Speeches-Statements.htm?dtl/25050/Prime_Ministers_address_to_UNESCO_April_10_2015.

MEA. 2015b. "Joint Statement on Climate Change between India and China during Prime Minister's Visit to China." *Ministry of External Affairs, Government of India*. May 15. Accessed August 9, 2020. https://mea.gov.in/bilateral-documents.htm?dtl/25238/.

MEA. 2015c. "Indo-German Joint Statement on Climate Change and Energy Technology Cooperation." *Ministry of External Affairs, Government of India*. October 5. Accessed August 10, 2020. https://mea.gov.in/bilateral-documents.htm?dtl/25884/IndoGerman_Joint_Statement_on_Climate_Change_and_Energy_Technology_Cooperation_October_05_2015.

MEA. 2015d. "India-Africa Cooperation in Science and Technology – Capacity Building." *Ministry of External Affairs, Government of India*. October 19. Accessed August 9, 2020. https://www.mea.gov.in/in-focus-article.htm?25947/IndiaAfrica.

MEA. 2017. "India-Germany Joint Statement during the Visit of Prime Minister to Germany." *Ministry of External Affairs, Government of India*. May 30. Accessed August 9, 2020. https://www.mea.gov.in/bilateral-documents.htm?dtl/28496/IndiaGermany.

MEA. 2018. "Address by External Affairs Minister at the International Solar Alliance Founding Conference." *Ministry of External Affairs, Government of India*. March 11. Accessed August 8, 2020. https://www.mea.gov.in/Speeches-Statements.htm?dtl/29601.

MEA. 2020. "India: UN Security Council Candidate." *Ministry of External Affairs, Government of India*. Accessed August 9, 2020. https://www.mea.gov.in/Images/amb1/INDIAUNSC.pdf.

Miller, Christopher P. 2020. "Soft Power and Biopower: Narendra Modi's "Double Discourse" Concerning Yoga for Climate Change and Self-Care". *Journal of Dharma Studies* 3 (1): 93–106.

MNRE. 2015. "Working Paper on International Solar Alliance (ISA)." *Ministry of New and Renewable Energy, Government of India*.

Modi, Narendra. 2010. *Convenient Action: Gujarat's Response to Challenges of Climate*. New Delhi: Macmillan India Limited.

Modi, Narendra. 2019. "PM's Remarks at Climate Action Summit 2019 During 74th Session of UNGA." *Narendra Modi*. September 23. Accessed

August 9, 2020. https://www.narendramodi.in/pm-modi-s-remarks-at-summit-on-climate-change-546575.

Modi, Narendra. 2015. "PM Authors "Convenient Action – Continuity for Change"." *Narendra Modi.* December 9. Accessed August 5, 2020. http://www.narendramodi.in/pm-authors-%E2%80%9Cconvenient-action-continuity-for-change%E2%80%9D-386189.

Mohan, Archis. 2015. "BJP Proposes 'panchamrit' to Nehruvian 'panchsheel'." *Business Standard.* April 4. Accessed August 8, 2020. https://www.business-standard.com/article/politics/bjp-proposes-panchamrit-to-nehruvian-panchsheel-115040300850_1.html.

Mohan, Kshitiz. 2017. "Make in India: Renewable Energy Sector Sees Fast Growth." *Business World.* March 16. Accessed August 9, 2020. http://www.businessworld.in/article/Make-In-India-Renewable-Energy-Sector-Sees-Fast-Growth/16-03-2017-114526/.

Mohan, Vishwa. 2015. "Obama-Modi Climate Deal: Unlike China, No Emission Target for India." *The Times of India.* January 26. Accessed August 8, 2020. https://timesofindia.indiatimes.com/home/environment/global-warming/Obama-Modi-climate-deal-Unlike-China-no-emission-target-for-India/articleshow/46016298.cms.

Mohan, Vishwa and Bagchi, Indrani. 2014. "After US-China Deal, India May Have to Reset Climate Goals." *The Times of India.* November 13. Accessed August 9, 2020. https://timesofindia.indiatimes.com/home/environment/pollution/After-US-China-deal-India-may-have-to-reset-climate-goals/articleshow/45130021.cms.

Mohapatra, Nalin Kumar. 2019. "Why the International Solar Alliance Is Geopolitically Significant." *Down to Earth.* April 19. Accessed August 9, 2020. https://www.downtoearth.org.in/blog/energy/why-the-international-solar-alliance-is-geopolitically-significant-64080.

Nagarajan, Ganesh. 2015. "India to Quadruple Renewable Capacity to 175 Gigawatts by 2022." *Bloomberg.* February 28. Accessed August 10, 2020. https://www.bloomberg.com/news/articles/2015-02-28/india-to-quadruple-renewable-capacity-to-175-gigawatts-by-2022.

Narain, Sunita. 2015. "Sunita Narain: Paris - The Endgame for Climate Justice." *Business Standard.* December 20. Accessed August 5, 2020. http://www.business-standard.com/article/opinion/sunita-narain-paris-the-endgame-for-climate-justice-115122000640_1.html.

Neuhauser, Alan. 2015. "U.S.-India Deals on Clean Energy, Nuclear Power Pave Way for Paris." *US News.* January 26. Accessed August 10, 2020. https://www.usnews.com/news/articles/2015/01/26/us-india-breakthrough-on-climate-change-nuclear-power-pave-way-for-paris.

PIB. 2019a. "Prime Minister Announces Coalition for Disaster Resilient Infrastructure at UN Climate Action Summit 2019." *Press Information Bureau.* September 24. Accessed August 9, 2020. https://pib.gov.in/PressReleaseIframePage.aspx?PRID=1586051.

PIB. 2019b. "Joint Statement Issued at the Conclusion of 29th BASIC Ministerial Meet on Climate Change." *Press Information Bureau, Government of India.* October 26. Accessed August 9, 2020. https://pib.gov.in/Pressreleaseshare. aspx?PRID=1589318.

PTI. 2015a. "India Will Set Climate Change Conference Agenda: Modi." *The Hindu.* April 14. Accessed August 8, 2020. https://www.thehindu.com/ news/national/india-will-set-climate-change-conference-agenda-modi/ article7101281.ece.

PTI. 2015b. "Paris Deal Window Dressing as India to Double Coal Production." *Financial Express.* December 15. Accessed August 9, 2020. https:// www.financialexpress.com/economy/paris-climate-change-summit-window-dressing-as-india-to-double-coal-production/179233/.

PTI. 2018. "India Willing to Take Lead in Combating Climate Change: Sushma Swaraj." *The Economic Times.* September 27. Accessed August 8, 2020. https://economictimes.indiatimes.com/news/politics-and-nation/ india-willing-to-take-lead-in-combating-climate-change-sushma-swaraj/ articleshow/65977912.cms?from=mdr.

PTI. 2019. "India "Walking the Talk" on Climate Change Commitments, Says Prakash Javadekar at COP 25." *The Economic Times.* December 10. Accessed August 9, 2020. https://economictimes.indiatimes.com/ news/politics-and-nation/india-walking-the-talk-on-climate-change-commitments-prakash-javadekar/articleshow/72460454.cms.

Purushothaman, Chithra. 2018. "India's Rising Stature as a Solar Power." *The Diplomat.* March 14. Accessed August 5, 2020. https://thediplomat. com/2018/03/indias-rising-stature-as-a-solar-power/.

Rajamani, Lavanya. 2016. "Ambition and Differentiation in the 2015 Paris Agreement: Interpretative Possibilities and Underlying Politics." *International and Comparative Law Quarterly* 65 (2): 493–514.

RNZ. 2017. "India to Help Pacific with Climate Change." *Radio New Zealand.* September 5. Accessed August 9, 2020. https://www.radionz.co.nz/ international/pacific-news/338705/india-to-help-pacific-with-climate-change. https://economictimes.indiatimes.com/industry/energy/power/international-solar-alliance-may-extend-its-membership/articleshow/63241695.cms.

Saran, Samir and Jones, Aled. 2017. *India's Climate Change Identity: Between Reality and Perception.* London: Palgrave Macmillan.

Saryal, Rajnish. 2018. "Climate Change Policy of India: Modifying the Environment". *South Asia Research* 38 (1): 1–19.

Saurabh. 2018. "India Offers $1 Billion to African Countries for Solar Projects." *Clean Technica.* March 20. Accessed August 9, 2020. https://cleantechnica. com/2018/03/20/india-offers-1-billion-african-countries-solar-projects/.

Sethi, Nitin. 2014. "India Cannot Afford a Climate Change Skeptic as Prime Minister". *Business Standard.* September 10. Accessed August 9, 2020. https://www.business-standard.com/article/opinion/india-cannot-afford-a-climate-change-skeptic-as-prime-minister-114091000128_1.html.

Sidhu, Waheguru Pal Singh. 2017. "'Vasudhaiva Kutumbakam' for the 21st Century." *Brookings*. May 22. Accessed August 9, 2020. https://www.brookings.edu/opinions/vasudhaiva-kutumbakam-for-the-21st-century__trashed/.

Tripathi, Bhaskar. 2018. "Budget 2018: India's Renewable Progress Slips; GST-Induced Losses, Import Duty on Solar Modules Threaten 2022 Target." *First Post*. January 25. Accessed August 9, 2020. https://www.firstpost.com/business/budget-2018-indias-renewable-progress-slips-gst-induced-losses-import-duty-on-solar-modules-threaten-2022-target-4319073.html.

UNFCCC. 2014. "Statement by Mr. Prakash Javadekar, Hon'ble Minister of State with Independent Charge for Environment, Forests & Climate Change." *United Nations Framework Convention on Climate Change*. December 9. Accessed August 10, 2020. https://unfccc.int/files/meetings/lima_dec_2014/statements/application/pdf/cop_20_hls_india.pdf.

Venugopal, Vasudha. 2016. "Manufacturing to Move Into 'Zero Defect, Zero Effect' Category." *The Economic Times*. January 21. Accessed August 9, 2020. https://economictimes.indiatimes.com/news/economy/policy/manufacturing-to-move-into-zero-defect-zero-effect-category/articleshow/50664212.cms?from=mdr.

VOA News. 2016. "India and Climate Change." *VOA News*. March 10. Accessed August 10, 2020. https://www.youtube.com/watch?v=ebpVByMnT6o.

Walt, Vivienne. 2015. "India's Modi Demands 'Climate Justice' at Paris Summit". *Fortune*. December 1. Accessed August 8, 2020. https://fortune.com/2015/11/30/narendra-modi-climate-justice/.

WRI. n.d. "Navigating the Paris Agreement Rulebook." *World Resources Institute*. Accessed August 8, 2020. https://www.wri.org/paris-rulebook.

6 Conclusions

An analysis of shifts in India's climate diplomacy positions

India's climate diplomacy has been heavily influenced by values, ideas, and norms that are social constructed over a period and influenced by material forces. The ideational drivers of the shifts in India's positions range from ethical to geopolitical, with socio-economic, technological, cultural, and ecological variables acting mostly as intervening variables, as explained in the previous chapters. Despite the apparent deficit in material and structural power that India possesses in relative terms, its rise to the position of a potential climate leader can largely be attributed to its ability to build and advance norms in the international arena. The image of a responsible global player is, undoubtedly, backed up by India's aspirations to be recognised as a power to reckon with and its efforts to develop material capabilities.

Before the 2007 Bali Summit, the formula narrative of India's climate diplomacy was primarily guided by concerns of equity and strategic autonomy (linked closely to the idea of sovereignty). Moral and political ideas had been equally influenced by the debate between environment and development, emphasis on poverty reduction, and more importantly, the North–South discourse – in which the Indian delegates typically upheld the "historical responsibility" principle. In this narrative, India's identity was largely dominated by that of a developing country that did not need to undertake climate action. Neither was it in a position to do so owing to socio-economic imperatives and the lack of financial and technological resources. (Geo)politically, India was averse to the idea of Western perspective on climate change and the hierarchical climate order, in which its position was regarded as expendable. When it comes to climate vulnerability, there existed much less understanding of it not only because of the lack of scientific assessment but also owing to the scepticism of climate science produced by some of the developed countries. The civilisational and cultural values have remained as underlying forces, but they did not direct the discourse of India's stated positions internationally.

In the run-up to the Copenhagen Summit and up until the Durban Summit, the international climate order had seen significant shifts, primarily guided by the emergence of the fast-growing economies of China, India, Brazil, and South Africa. India's climate diplomacy positions during this period have been derived as much from geopolitical interests as equity and other ethical concerns. Although India continued to uphold the principle of CBDR-RC and press for more financial and technological transfer, it began to recognise the need for showing enhanced commitment at the international level, as climate change had emerged as a critical issue of global governance. The global transition was also marked by the emergence of the idea of shared responsibility wherein all countries must act (with conditionality such as finance and technology). With this, the legal divide between the developed and developing countries under the Kyoto Protocol was watered down to some extent. This also led to the enhanced emphasis on issues such as transparency (of climate action), and "internationalisation" of actions through greater international cooperation – something that the Indian delegation began to concede to progressively. India started to give into the developed countries' demand for greater level of "mutual scrutiny of domestic actions", which was ultimately formalised at the Copenhagen Summit as a part of various other packages that were agreed upon, including the Green Climate Fund (GCF). Most importantly, the shift in the global climate architecture itself – from a top-down model to a bottom-up one – also gave impetus to India's international positioning based on its national interests and priorities.

The period from 2009 to 2015 was marked by geopolitical and geoeconomic flux. As the rich countries were struggling to come out of the financial crisis, the emerging economies of China and India grew economically with minor setbacks only. The power shifts were imminent with the growing influence of the emerging economies in global governance, including on the issue of climate change. China and India's growing GHG emissions also led to a shift in the international community's perception of these countries. This impinged upon India's principle or strategy of "per capita plus" too that led India to go beyond the per capita emissions principle. Similarly, the geopolitical realignments in the form of the emergence of the G-2 and the 2014 US–China joint communiqué also played a role in conditioning India's positions. For instance, despite cooperation with China in the realm of climate change, especially as a part of the BASIC grouping, the solidarity between them on issues such as equity, financial and technological assistance, and MRV began to crumble, as China began to gradually position itself alongside the US and the EU on several fronts (such as mitigation commitments).

During 2009–2015 (more so until 2013), in terms of diagnostic framing, India did not waver too much – continuing to uphold "historical responsibility", thereby also emphasising the importance of the second commitment period of the Kyoto Protocol (2012–2020). As a corollary, in terms of prognostic framing, India, along with the other developing countries, urged the rich countries to fulfil the commitments of the Kyoto Protocol and ramp up climate finance at the same time. However, with the realisation that the transfer of technology and finances from the developed countries to the developing countries was going to be difficult to materialise, the ethical issues of equity and climate justice began to be used more as an argument for India's non-willingness to take on more commitments rather than as a demand. Therefore, India's position on finance and technology began to be diluted.

Similarly, "equity", which was initially used to differentiate between the developed and developing countries (based on emissions reduction commitments), was largely considered by India as the bedrock of distribution of the available carbon space (or development space) equitably. This is linked to both its relatively low per capita emissions (based on current emissions and population) and historical responsibility. After the Bali Action Plan, once India realised that the notion of carbon space was not getting much traction (owing to the developed countries' opposition to it), it shifted the argument towards equity expressed in terms of space for "future" sustainable development. As a part of this shift, it also agreed to the concept of NAMAs, albeit supported by international finance. In short, the principle of historical responsibility was weakened in the negotiations. However, countries could not agree upon the parameters for the operationalisation of equity, which later on led to the adoption of equity in terms of a qualitative notion of "fairness" – delineating actions that a country could adopt based on national circumstances and domestic capabilities.[1]

As a part of India's diagnostic framing of climate change, it began to see climate action as an opportunity – with co-benefits. Therefore, one of the major shifts (prompted also by the launch of the National Action Plan on Climate Change in 2008) was interlinkages drawn between climate and energy goals, with a focus on renewable energy. In effect, the "development" (socio-economic) frame was further modified to include issues of energy security and sustainable growth. Moreover, the industry and businesses within the country helped shape the debates surrounding flexibility mechanisms such as the CDM and the benefits of measures such as energy efficiency for cost and operational effectiveness. It also became a question of geopolitical posturing for India, as explained earlier. Accordingly, as a part of its prognostic framing, one of the first steps that India took internationally was to declare an emissions intensity (of its GDP) reduction target at the 2009 Copenhagen Summit, which India seems to be in a good position to achieve under the current trends and projections. However, India continued to press for financial and technological assistance for the developing countries, under a bottom-up framework of climate governance, in which countries such as India would also commit to certain goals (with the condition that international scrutiny of its domestic actions would be based on "consultation" and "analysis").

As far as the motivational framing is concerned, India was not only under international pressure to be more flexible, but its geopolitical ambition also played a big part in moulding its positions to become more accommodative and compromising on some of the key demands of both the developed and least developed countries. With the growing scientific consensus over the impact of climate change on the increasing frequency and intensity of disasters and the consequent losses incurred by the most vulnerable countries, India's position in the international climate order became more tenuous than ever before. Therefore, concerns over isolation by the rest of the developing world, including its own neighbours (such as the Maldives and Bangladesh), coupled with the growing fissures among the BASIC countries led India to pave its own pathway in the negotiations. This pathway is as dependent on its own national interests as well as the interests of the other parties who expected India to make concessions on certain issues such as MRV (scrutiny) and the operationalisation of equity for a strong post-2020 climate regime. However, it must also be added here that India did not budge on its goal of doubling coal production and achieving self-sufficiency in the coal sector by 2024, despite severe opposition by the rest of the international community. Therefore, India's domestic climate/energy policy could not be entirely held hostage to international pressure, as it grapples with energy poverty.

India's motivational framing would be incomplete without referring to the aspirational facets of India's foreign policy. India's readiness to be a part of the solution and not be seen as a "bad boy" began to supersede other dominant framings of climate change. Even though the gradual shifts in positions that were spearheaded by Jairam Ramesh saw a brief lull when he was shifted out of the Ministry of Environment and Forests, Prime Minister (PM) Narendra Modi's ascendancy to power ensured the return of the relevance of India's image and reputation at the international level as a responsible player. As the study points out, India's positions were purportedly influenced by not only its desire to be a permanent member of the United Nations Security Council (UNSC) but also its apprehension over being isolated by the other countries. In effect, in the run-up to the Paris Summit, the frame of international stature had a bearing on India's red lines. In this context, the launch of the International Solar Alliance, a treaty-based organisation headquartered in India, is noteworthy.

Another addition to India's formula narrative and motivational framing, particularly after the Modi administration took over the reins of power, was the reinforcement of India's civilisational identity, embedded with the ethos of sustainability. Even though poverty and energy crisis top the list of priorities for any Indian administration, enhancing India's image at the international level through its age-old commitment to the environment to deflect and/or counter debates on its stance became more and more evident. Furthermore, at the domestic level, India's conceptualisation of national interests, based on its ecological vulnerability and the co-benefits of climate action, helped bring forth the relevance of cooperation at various levels (multilateral, bilateral etc.) with other countries and groups of countries. Even though it is difficult to say whether bilateral relations have influenced shifts in India's multilateral positioning on climate change, it cannot be denied that a pragmatic and results-oriented approach became increasingly relevant in the context of India's climate diplomacy in general. In the same vein, it might be difficult to gauge the influence of India's increasing ecological vulnerability on its positions in the negotiations, but its expanding science and technology base (including in climate science and modelling, and renewable energy) has given it more confidence to engage in various climate change–related forums. It has also been a motivational factor in spurring many bilateral initiatives in sectors such as energy, clean air, transport, resilience etc.

To sum up, the period 2009–2015 marked a shift both internationally and domestically for India. Its formula narrative reflected these changing realities. From South–South cooperation, India's close cooperation and coordination with the developed countries and other emerging economies/large developing countries grew. The question of finances and technological support began to be used as a shield rather than a sword.[2] On the one hand, India focussed on the pre-2020 commitments (second commitment period of the Kyoto Protocol) and on the other, it started to build channels with several countries to enter bilateral agreements on specific sectors that could gain from climate cooperation. The geopolitical ideas dominated the discourse as India's efforts to portray itself as a responsible player became apparent. These ideas have been influenced by socio-economic, technology, and cultural factors. The ethical concerns were not entirely eclipsed as they continued to be a part of the narrative, but their decisiveness was blunted by ideational and material shifts, as India showed signs of flexibility. These concerns began to be regarded as opportunity costs in India's broader foreign policy strategy and, therefore, it was keen on being seen at the centre of the signing of the Paris Agreement in 2015 rather than in the periphery. However, the ideas of sovereignty and strategic autonomy continued to be upheld in many cases, including on the questions of coal and MRV, as well as when the US–China climate agreement occurred and expectations were high that the US would be able to get India also to sign a similar agreement.

The period between 2015 and 2019 was tumultuous in many respects. To start with, President Trump took over the reins of power in the US. This shift had its reverberations among the international community. From the EU, China, and India attempting to fill the void left by President Trump's announcement to withdraw from the 2015 Paris Agreement, to the crumbling of the global multilateral order itself, 2015–2019 period saw a significant decline in the upbeat mood that the world had attained after signing the Paris Agreement. Pushbacks from countries such as Brazil, Australia, and Saudi Arabia towards the Paris Agreement and its provisions further have complicated the process. The second commitment period of the Kyoto Protocol never came into effect, with Canada repudiating it completely and others such as Japan and Russia refusing to adopt further commitments under its framework. Therefore, India was caught in a dilemma – between its aspirational goals (tied to foreign policy) and its long-standing demand of fulfilling the pre-2020 commitments by the rich countries. In any case, the gradual trend of linking India's domestic actions (in sectors such as energy) with its international climate diplomacy position did not see any change. Domestically, India has been focussing on both enhancing its knowledge in the spheres of climate science, mitigation, and resilience. The role of non-state and sub-state actors (including the media) in highlighting India's increasing climate vulnerabilities and gaping holes in climate policy as well as calling for more action by the government has, however, bolstered.

The diagnostic, prognostic, and motivational framing of climate change by India did not alter drastically during 2015–2019. Institutional and bureaucratic changes in the domestic climate policy architecture may have conditioned its climate diplomacy positions, but these factors were not decisive enough, as the study also points out. India's bilateral and multilateral push did see an upward trend (in particular, with parties such as the EU), but the overall sentiment was sluggish with President Trump naming China and India in his Paris Agreement withdrawal speech. India repeatedly vowed its commitment to the Paris Agreement as well as took the moral high-ground in meetings such as the 2019 Climate Action Summit. Diagnostically, climate change is still a development issue, with sustainable development and co-benefits at the centre of its positioning. At the same time, the focus on lifestyle and behaviour, particularly in the rich countries, became a target area for PM Modi. Although this is not new to India's formula narrative, the framing of climate change as a behavioural issue within the solutions framework had not sufficiently been drawn upon. For example, India's soft power such as yoga came to be discussed more often as a solution to climate change. The discourse shifted to what many would say – what India could offer to the rest of the world.

Apart from the above-mentioned prognostic framing, India began to work towards tapping into avenues of cooperation with the other developing countries in areas such as solar energy. The leadership began to focus on working with the "willing" in specific areas. At the same time, India used the platform of LMDC to demand equity and CBDR-RC in the post-2020 regime in all aspects (including market mechanisms that are yet to be finalised at the time of writing this work). As far as motivational framing is concerned, yet again, it is a mix of geopolitical, ethical, socio-economic, technology, cultural, and ecological ideas, which constitutes India's climate diplomacy. As much as India insists on the pre-2020 commitments (mitigation and finance) by the rich countries, it is equally conscious about achieving the NDCs under the Paris Agreement. Even while demanding the CDM credits to be carried forward (which many countries are opposed to), it has been providing thrust to climate solutions by reemphasising the importance of the solar alliance and, simultaneously, engendering new initiatives such as the Coalition for Disaster Resilient Infrastructure (CDRI). This assumes greater significance as it is a clear reflection of India's inability to either compel the rich countries to accede to the developing countries' demands on the loss and damage mechanism or even have sufficient access to any type of adaptation finances (as they are mostly allocated to the LDCs and island countries). Additionally, as these moves explain, India's self-cognisance of its worsening climate scenario is evident.

While one might argue that India's concerns with regard to equity have slowly lost steam as years passed, this may not be entirely true. As contended by some experts, India's climate diplomacy has not changed remarkably. India's stance has been clear – for sustainable consumption and development, poverty eradication, climate justice, and CBDR-RC. India has never changed its stance on the diagnostic framing as far as the causes of climate change are concerned. It has always blamed the developed world for causing climate change, while portraying itself as suffering the brunt of it. As Indrajit Bose states, "India's position has been – we need our policy space to grow; do not trample upon it". Its attempts to thwart the obstructionist image imposed on it by the Western media is also noteworthy in this context. As Bose comments, as of now, the developing countries cumulatively have much higher commitments as compared to the developed countries. The latter have also reneged on various commitments under the Kyoto Protocol. Under these circumstances, "they have no right to say that India has been obstructing the negotiations".[3]

Patterns of emerging economies' climate diplomacy positions

In this section, the identifiable common patterns in the emerging economies' climate diplomacy positions are explicated. The countries in focus are – Brazil, South Africa, India, and China. The patterns are analysed by looking into the five ideational categories – ethical/moral, ecological/environmental, cultural/civilisational, socio-economic and technology, and political/geopolitical. Both the domestic and international factors are taken into consideration in this context.

If there is one set of ideas that are most responsible for bringing the BASIC countries together, it is the ethical/moral ones. Notions of "historical responsibility" of the developed countries for creating the climate change problem and, hence, their responsibility to find solutions and carry the burden of implementing them have dominated their common position. However, from being extremely stringent on these issues, these countries have been more flexible in their approach towards the operationalisation of equity in recent years, particularly in the run-up to the Paris Summit. After the Copenhagen Summit, when the four countries showed "solidarity" in their opposition to the developed countries' demands, Brazil and South Africa moved gradually towards a greater understanding with the developed world on the future climate regime that would be "applicable to all". China and India bargained (to protect their interests) for a longer period of time, which also took a turn when China acquiesced with the developed world's demands (for example, the 2014 US–China agreement). One could argue that India withstood the pressure up until the Paris Summit, after which it also realised the pulse of the negotiations, which prompted it to change its negotiation tactics and compromise on some of its red lines. The fact is that, today, moral and ethical concerns have largely become an argument in the negotiations space and may not have as much actionable or instrumental value. Yet, the global climate order is built upon the UNFCCC and various other historical narratives/frameworks that are derived out of these debates between the developed and developing countries. They cannot be wished away anytime soon – especially when the question of inequities and injustice have not been addressed by the developed world (in the eyes of the developing world).

One of the inferences drawn from the analysis is that the idea of ecological vulnerability has rarely influenced climate diplomacy positions of the emerging economies internationally, except in the formative years of the negotiations. Despite being recognised as highly climate-vulnerable countries, except South Africa, other emerging economies have tended to use vulnerability as a peripheral factor and not a core one. South Africa, owing to its regional African identity that tends to reinforce this aspect, has also been mellower in comparison to other countries of the region. Yet, it has been more interested in adaptation-related discussions than the other emerging economies. At the same time, the ecological vulnerabilities faced by the developing countries, particularly in relation to disasters (and associated losses) have brought these countries together to push for a stronger loss and damage mechanism (if not for themselves, but for other developing countries). China has also been using its growing geoeconomic capabilities to invest way more than the other emerging economies in other developing countries' climate resilience projects and programmes (Liqiang 2019). However, countries such as Brazil and South Africa have been most concerned about the effects of climate mitigation on their forest and coal sectors, which are major contributors to their economic growth.

The cultural and civilisational values have been more prominent in the climate diplomacy positions of the emerging economies during 2009–2019. From South Africa's *Indaba* and China's "ecological civilisation" to India's *Vasudhaiva Kutumbakam*, cultural ideas propel dynamics of multilateralism and international cooperation as well as climate leadership. Although Brazil has a rich cultural history in terms of the Amazon (indigenous cultures and their ecological practices etc.), it has not actively used it to project this identity. In fact, even under the most "progressive" (environmentally) Brazilian leaderships (such as Presidents Lula and Rousseff), these cultures have come under attack (Branford and Torres 2017). While the post-colonial identities have played a role in pursuing the anti-colonial (and in turn, anti-West) agenda even in the formative years of the climate change negotiations, these began to be used more as a means of showcasing their leadership rather than merely imprinting a rhetorical stance. In particular, countries such as China and India with their civilisational history felt it apt to bring about the relevance of these ideas and values in shaping the global climate order as well.

As far as socio-economic and technology ideas are concerned, these are intricately intertwined with the other categories. The emerging economies have moved from mere preoccupation with the notions of development space and poverty eradication to aspirations of expanding clean energy base (technological leadership) and even just transition (particularly for South Africa). This shift has been most evident during 2009–2015 when countries such as China and India that had held on to their demand for more flexibility in the IPR regime in the realm of green technology transfer started to focus more on building their own S&T base. China, in particular, has emerged as the global leader in clean energy and left even most of the developed countries behind in this race. While Brazil has been keen on addressing the problem of deforestation in the Amazon through finances from the developed world, South Africa has been insisting on financial support to ensure just transition in a country that is heavily dependent on coal. All these countries are geared towards treating climate change as an opportunity to bring reforms in certain sectors such as energy and transport. The co-benefits approach has been at the centre of India's strategy of tying climate goals to that of energy security.

Last but not the least, political/geopolitical interests and ideas have been instrumental to the emerging economies' climate diplomacy positions. Domestically, the political and institutional changes have brought about several nuanced changes in their positions. It is perhaps more stark in the case of Brazil, which recently witnessed the rise of President Bolsonaro. In the case of other countries, there has largely been continuity. The countries, as already explained, have shown solidarity and also been portrayed as bridging nations between the developed and developing countries on various occasions. There have been fissures, but at least on issues related to the principles of equity and climate justice, they still remain united. Geopolitically, a group such as BASIC was successful in resisting pressure from the developed countries on several occasions, including at the Copenhagen Summit. For these countries, climate change was largely a foreign policy issue for a long time, until countries began to introduce and institutionalise policies and agencies at the domestic level – not only to act upon climate change but also for international consumption.

The emerging economies have steadfastly espoused a position that would not compromise their sovereignty – based on their notions of non-interference in their domestic affairs and the principle of inalienable right to utilise their resources to eradicate poverty. A country like India has also, in addition, put emphasis on strategic autonomy, derived from its foreign policy. These ideas have principally influenced their positions on various aspects of the negotiations – mitigation, transparency, and so on. From Brazil's right to use its forest resources to South Africa's right to exploit its coal reserves, sovereignty has a bearing on their choices. China and India, on the other hand, have been more vociferous on issues related to transparency (even during 2009–2019). The emerging economies have multiple identities (geopolitically), but except South Africa, the others do not share a regional climate identity, as far as common positions on climate change are concerned. Similarly, China is far ahead in terms of economic growth, GHG emissions, and geopolitical influence, which places it in a different league in comparison to the other emerging economies.

For all the four emerging economies, recognition as a responsible player and a global climate leader had assumed significance in the run-up to the Paris Summit. This was particularly important for China and India, which had been viewed as obstructionists at and after the Copenhagen Summit by the developed countries and even some of the LDCs and island nations. However, international pressure had differentiated influence on the four countries' positions. For China, it was a very important factor, which is why the issue of climate change was elevated to the level of the State Council – the highest deliberative body in the country.[4] It does not seem to have been as big a factor in the other cases, presumably due to their lower emissions and economic profiles. In South Africa's case, the pressure was evidently built not only by the developed countries but also the LDCs within Africa who have been most keen on more stringent and urgent climate action, which is linked to greater access to financial and technological assistance as well.

As much as the evolving geopolitics affects the global climate order, climate change also can be considered to be shaping geopolitics. New alignments and partnerships surrounding climate change have emerged in recent years. Emerging economies such as China have been using climate change to advance their geopolitical interests, as steps taken by the country's leadership in tackling the issue may help enhance its comprehensive national power or *zonghe gouli*. For India, the debate on whether climate change can drive international politics still continues. Therefore, there may still be scepticism about whether India's actions on climate change could actually change the balance of equations in other areas. However, the leaderships of these countries are largely pragmatic about their approaches and positions, and protective about their bilateral and multilateral interests while dealing with this issue. Keeping in mind the reputational risk associated with inaction, they vow to contribute to global governance as a responsible player, even while framing climate concerns with their domestic interests.[5]

This study brings to fore the close coordination of ideational and material factors in the emerging economies' climate diplomacy positions during 2009–2019. As established, ideas have certainly been more influential in the building of these narratives than material forces. However, this does not imply that material forces are insignificant. In fact, the emerging economies' rise in the international system and their aggregate bargaining powers have certainly helped them wield more influence and consolidate their position within the international climate order. However, as clearly expressed, it is through the changing perceptions of certain ideas – ethical, cultural, ecological, socio-economic, and technology and (geo)political – that these countries' climate diplomacy positions can best be studied. Not only have these countries' framing of climate change and action changed over a period of time, their perceptions surrounding their role and responsibility on the issue have evolved too.

The emerging economies have been able to bring about changes in the climate order itself. For example, the bottom-up model of climate governance was as much a handiwork of the deliberations among the emerging economies, as between them and the US. Despite relatively lesser material capabilities, a country like South Africa played an important role in providing fillip to the negotiations in Durban, which paved the way for the Paris Agreement. Similarly, India launched the International Solar Alliance – founded on the ideas of climate justice as well as its increasing material capabilities in the solar sector. China has been able to present itself as a global climate leader, and is now well-placed to steer the climate regime in the post-2020 era, along with the other 'willing' countries. Backed by its decreasing deforestation rates and willingness to cooperate on reducing further forest degradation, Brazil is placed alongside the other proactive climate players too (at least till 2019).

Through various steps, the emerging economies have been able to dispose of the negative image (even if, it is unfair) they acquired at the Copenhagen Summit. However, there are still many gaps as far as the international image vis-à-vis domestic actions are concerned. There is ample scope for research on this dissonance between the international and domestic scenarios. Some of these aspects have been covered in the study, but the level of influence that domestic factors and national-level policies have on international positioning (particularly in climate talks) needs to be studied further. Even though they may seem to influence and reinforce each other at the outset, there are areas in which they may not readily coalesce; and therefore, these dynamics could be looked into more deeply. Similarly, one of the major gaps in research is bilateral climate diplomacy and what effects it has or could potentially have on the multilateral process. Again, an effort has been made in this study to explore bilateral processes between some countries, but more research on this question could unravel several interesting aspects of the twenty-first-century climate diplomacy, especially in the post-2020 context. Considering that the Paris Agreement tries to bring more actors, tools, and actions on the table, the emerging economies would be expected to have a much bigger role in shaping the future of the international climate order through their norms and ideas.

Notes

1 The author's interview with R. R. Rashmi, Distinguished Fellow and Programme Director, Earth Science and Climate Change, The Energy and Resources Institute; and former India's principal negotiator for climate change negotiations under the UN Framework Convention on Climate Change; and Special Secretary in the Ministry of Environment, Forest & Climate Change in the Government of India via Microsoft Teams on August 17, 2020.
2 The author's interview with Navroz Dubash, Professor, Centre for Policy Research (CPR), New Delhi, via Zoom on August 10, 2020.
3 The author's interview with Indrajit Bose, senior researcher, Third World Network, via Skype on July 8, 2020.
4 The author's interview with Navroz Dubash.
5 The author's interview with R. R. Rashmi.

Bibliography

Branford, Sue and Torres, Maurício. 2017. "Indigenous Groups, Amazon's Best Land Stewards, under Federal Attack." *Mongabay*. Accessed August 2, 2020. https://news.mongabay.com/2017/04/indigenous-groups-amazons-best-land-stewards-under-federal-attack/.

Liqiang, Hou. 2019. "China Helps Developing Countries Tackle Climate Change." *China Daily*. December 13. Accessed August 2, 2020. https://www.chinadaily.com.cn/a/201912/13/WS5df34ac1a310cf3e3557e07f.html.

Index

Note: **Bold** page numbers refer to tables; *italic* page numbers refer to figures and page numbers followed by "n" denote endnotes.

Printed in the United States
by Baker & Taylor Publisher Services